Video training courses are available on
in the James Martin ADVANCED TECHNO...
Deltak Inc., East/West Technological Center, 1751 West Diehl Road,
Naperville, Ill. 60566 (Tel: 312-369-3000).

Books On Interactive Systems	Books On Distributed Processing	Books On Teleprocessing	Books On Telecommunications
VIDEOTEX	PRINCIPLES OF DISTRIBUTED PROCESSING	INTRODUCTION TO TELEPROCESSING	TELEMATIC SOCIETY
DESIGN OF MAN–COMPUTER DIALOGUES	COMPUTER NETWORKS AND DISTRIBUTED PROCESSING	INTRODUCTION TO COMPUTER NETWORKS	TELE-COMMUNICATIONS AND THE COMPUTER (second edition)
PROGRAMMING REAL-TIME COMPUTER SYSTEMS	DESIGN AND STRATEGY FOR DISTRIBUTED PROCESSING	TELEPROCESSING NETWORK ORGANIZATION	COMMUNICATIONS SATELLITE SYSTEMS
DESIGN OF REAL-TIME COMPUTER SYSTEMS	DISTRIBUTED FILE AND DATA-BASE DESIGN	SYSTEMS ANALYSIS FOR DATA TRANSMISSION	FUTURE DEVELOPMENTS IN TELE-COMMUNICATIONS (second edition)

DIAGRAMMING TECHNIQUES
FOR ANALYSTS
AND PROGRAMMERS

A ——————— BOOK

DIAGRAMMING

FOR

AND

TECHNIQUES
ANALYSTS
PROGRAMMERS

JAMES MARTIN
CARMA McCLURE

PRENTICE-HALL, INC., Englewood Cliffs, New Jersey 07632

Library of Congress Cataloging in Publication Data

MARTIN, JAMES (date)
   Diagramming techniques for analysts and programmers.

   Includes bibliographies and index.
   1. System analysis. 2. Electronic digital computers—
Programming. 3. Flow charts. I. McClure, Carma L.
II. Title.
T57.6.M3484 1984   001.64'23   84-17262
ISBN 0-13-208794-4

Editorial/production supervision: *Kathryn Gollin Marshak*
   and *Linda Mihatov*
Jacket design: *Whitman Studios, Inc.*
Manufacturing buyer: *Gordon Osbourne*

Diagramming Techniques for Analysts and Programmers
James Martin and Carma McClure

Printed in the United States of America

10 9 8 7 6 5 4 3 2

ISBN 0-13-208794-4

PRENTICE-HALL INTERNATIONAL, INC., *London*
PRENTICE-HALL OF AUSTRALIA PTY. LIMITED, *Sydney*
EDITORA PRENTICE-HALL DO BRASIL, LTDA., *Rio de Janeiro*
PRENTICE-HALL CANADA INC., *Toronto*
PRENTICE-HALL OF INDIA PRIVATE LIMITED, *New Delhi*
PRENTICE-HALL OF JAPAN, INC., *Tokyo*
PRENTICE-HALL OF SOUTHEAST ASIA PTE. LTD., *Singapore*
WHITEHALL BOOKS LIMITED, *Wellington, New Zealand*

TO CONSTANCE SUE

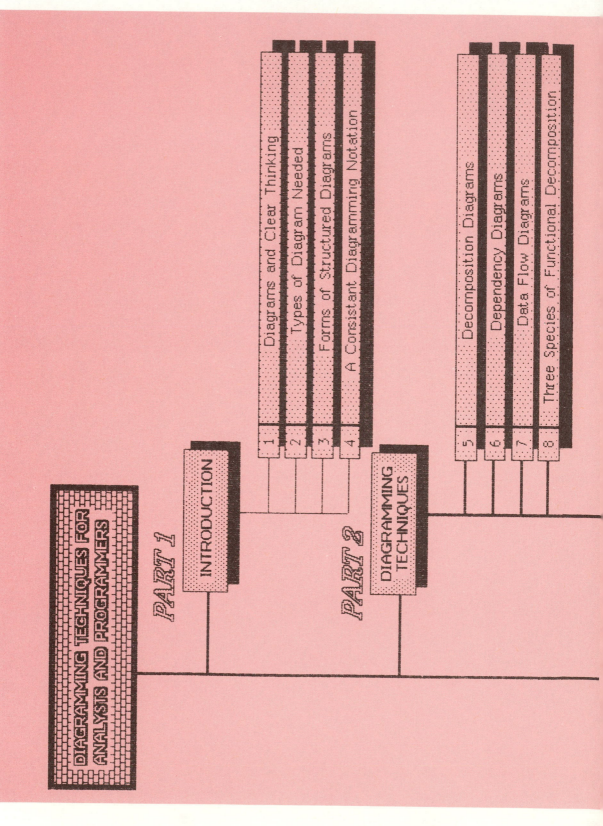

DIAGRAMING TECHNIQUES FOR ANALYSIS AND PROGRAMMERS

PART 1

INTRODUCTION

1   Diagrams and Clear Thinking
2   Types of Diagram Needed
3   Forms of Structured Diagrams
4   A Consistant Diagramming Notation

PART 2

DIAGRAMMING TECHNIQUES

5   Decomposition Diagrams
6   Dependency Diagrams
7   Data Flow Diagrams
8   Three Species of Functional Decomposition

CONTENTS

PART  **CONCLUSION**

PREFACE

The structured revolution has substantially changed systems analysis, design, and programming. But the structured revolution will be perceived as paving the way to a much grander revolution—CASA/CAP—Computer Aided Systems Analysis and Computer Aided Programming. The system designer will create his own designs at a graphics screen, and a computer will generate code from them.

There are, however, many incompatible diagramming techniques. Incompatible techniques can often be found in the design of a single system.

Which of these techniques are best? Do they need to be changed for computer-aided design? Can an integrated set of techniques be found to form the basis of the designer's workbench of the future?

This book reviews the diagramming techniques in common use and describes an integrated approach necessary for the most effective computer-aided design. It is a tutorial on diagramming which we believe should be basic knowledge for every analyst and programmer. It is aimed at the mass of computer professionals who need to clarify thinking and communication with the best diagrams. It is also aimed at DP executives who should be preparing for the CASA/CAP revolution which will sweep through the DP profession like a forest fire.

The philosophy of structured techniques pervades the book. A vital aspect of using structured techniques is to use the best types of diagrams. A diagram is worth a thousand words (and much more with some of the specifications we have examined). Many aspects of complex system design are much better stated in diagrams than in words.

The advocates of different techniques tend to make conflicting claims. In attending courses on the techniques, we found no course which represented the whole body of knowledge about structured techniques and diagrams which we think an analyst, designer, programmer, or DP manager ought to have today. This book addresses that body of knowledge, attempting to describe the techniques tutorially and set them into perspective.

ACKNOWLEDGMENTS We owe much to the professionals in James Martin Associates and Database Design Inc. for their comments on using and improving the diagramming techniques in practice when designing complex systems. In particular, the genius of Ian Palmer and Al Hershey is reflected in our writing.

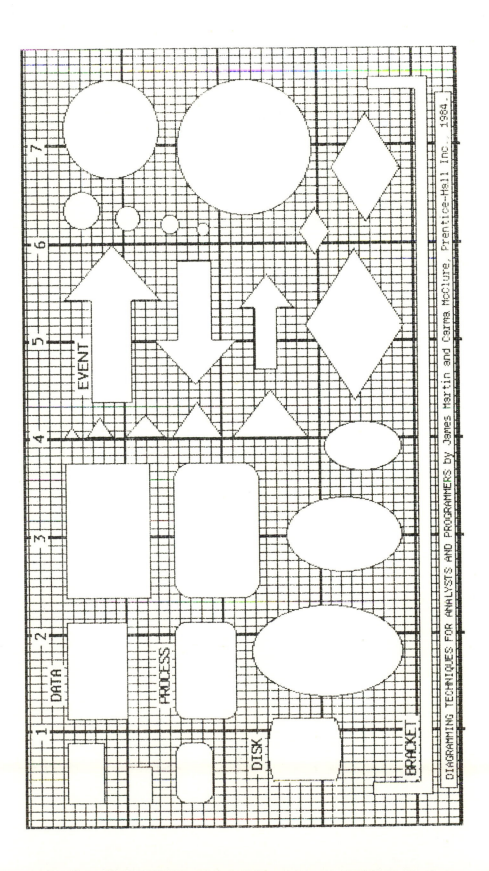

DIAGRAMMING TECHNIQUES FOR ANALYSTS AND PROGRAMMERS by James Martin and Carma McClure. Prentice-Hall Inc. 1984.

1 DIAGRAMS AND CLEAR THINKING

A FORM OF LANGUAGE   Good, clear diagrams play an essential part in designing complex systems and developing programs. Philosophers have often stated that what we are capable of thinking depends on the language we use for thinking. When mankind used only Roman numerals, ordinary people could not multiply or divide. That capability spread when Arabic numbers became widely used. The diagrams we draw of complex processes are a form of language. With computers we may want to create processes more complex than those which we would perform manually. Appropriate diagrams help us to visualize and invent those processes.

If only one person is developing a system design or program, the diagrams that he uses are an aid to clear thinking.* A poor choice of diagramming technique can inhibit his thinking. A good choice can speed up his work and improve the quality of the results.

When a number of people work on a system or program, the diagrams are an essential communication tool. A formal diagramming technique is needed to enable the developers to interchange ideas and to make their separate components fit together with precision.

When systems are modified, clear diagrams are an essential aid to maintenance. They make it possible for a new team to understand how the programs work and to design changes. When a change is made, it often affects other parts of the program. Clear diagrams of the program structure enable maintenance programmers to understand the consequential effects of changes they make. When

---

*The author has given much thought to the problem of avoiding words with a sexist connotation. It is possible to avoid words such as "man" and "manpower," but to avoid the use of "he," "his," and "him" makes sentences clumsy. In this book whenever these words appear, please assume that the meaning is "he or she," "his or her," and "him or her." They should be regarded as *neuter* words.

debugging, clear diagrams are again a highly valuable tool for understanding how the programs ought to work, and tracking down what might be wrong.

Diagramming, then, is a language, essential both for clear thinking and for human communication. An enterprise needs standards for data processing diagrams, just as it has standards for engineering drawings.

THE CASA/CAD REVOLUTION

Today, there is a new and very important reason for diagramming standards being well thought out. The job of systems analysts and builders is undergoing revolutionary change. It is evolving from a pencil-and-paper activity to an activity of computer-aided design. This change will enormously improve the productivity of systems builders and increase the quality of the systems they build.

Architects, engineers, and circuit designers have tools with which they can draw and manipulate diagrams on a computer screen. It is perhaps surprising that so many systems analysts and programmers still draw diagrams by hand. We have CAI, CAD, CAM (computer-aided instruction, computer-aided design, and computer-aided manufacturing). We need CASA and CAP (computer-aided systems analysis and computer-aided programming).

Interactive diagramming on a computer screen has major advantages. It speeds up the process greatly. It enforces standards. The computer may apply many checks to what is being created. It can automate the documentation process. The computer enforces discipline and permits types of cross-checking, calculation, and validity checks that human beings often do not apply. The large three-ring binders of specifications that have been typical are full of errors, inconsistencies, and omissions. Interactive graphics design with computers can eliminate much of the sloppiness, and replace volumes of text with powerful, computable symbolic design. Data processing installations everywhere should be making themselves ready for the CASA/CAD revolution.

Some diagramming techniques are more appropriate than others for automation. Automation of diagramming should lead to *automated checking of specifications* and *automatic generation of program code*. Many of the diagramming techniques of the past are not a sound basis for computerized design. They are too casual, unstructured, and cannot represent some of the necessary constructs.

BETTER THAN A THOUSAND WORDS

Given an appropriate diagramming technique, it is much easier to describe complex activities and procedures in diagrams than in text. A picture can be much better than a thousand words because it is concise, precise, and clear. It does not allow the sloppiness and woolly thinking that are common in text specifications.

The reader might glance at Fig. 24.13, at the end of the book, for example. The information conveyed in this decision tree would be lengthy and clumsy to express in text. Indeed, all of the diagrams at the end of Chapter 24 convey information with greater brevity and precision than does text.

Mathematics is the preeminent language of precision. However, it would be difficult to describe a large road map in mathematics. If we succeeded in doing so, the road map would still be more useful than mathematics is to most people. Data processing is complex and needs road maps. We need to be able to follow the lines, examine the junctions, and read the words on the diagrams.

Some diagrams, however, have a mathematical basis. With mathematics, we may state axioms that the diagrams must obey. A workstation may beep at a designer whenever he violates one of the axioms. The mathematically based structure that emerges may be cross-checked in various ways and may be the basis of automatic code generation.

THE NEED FOR FORMALITY

Architects, surveyors, and persons designing machine parts have *formal* techniques for diagramming which they *must* follow. Systems analysis and program design have even greater need for clear diagrams because these activities are more complex and the work of different people must interlock in intricate ways. There tends, however, to be less formality in programming as yet, perhaps because it is a young discipline full of brilliant people who want to make up their own rules.

One of the reasons why building and maintaining software systems is so expensive and error-prone is the difficulty that we have in clearly communicating our ideas to one another. Whether we are reading a functional specification, a program design, or a program listing, we often experience difficulty understanding precisely what its author is telling us. Whenever we must rely on our own interpretation of the meaning, the chance of a misunderstanding leading to program errors is very great.

The larger the team, the greater the need for precision in diagramming. It is difficult or impossible for members of a large team to understand in detail the work of the others. Instead, each team member should be familiar with an overview of the system and see where his component fits into it. He should be able to develop his component with as little ongoing interchange with the rest of the team as possible. He has clear, precisely defined and diagrammed interfaces with the work of the others. When one programmer changes his design it should not affect the designs of the other programmers, unless this is unavoidable. The interfaces between the work of different programmers need to be unchanging. To achieve this needs high-precision techniques for designing the overall structure of the system.

CHANGING METHODS

Diagramming techniques in computing are still evolving. This is necessary because when we examine techniques in common use today, many of them have *serious* deficiencies. Flowcharts are falling out of use because they do not give a structured view of a program. Some of the early structured diagramming techniques need replacing because they fail to represent some important constructs. We are inventing more rigorous methods for creating better specifications. Vast improvements are needed and are clearly possible in the specification process. These improvements bring new diagramming methods.

One of the problems with computing is that it is so easy to make a mess. Very tidy thinking is needed about the complex logic or a rat's nest results. Today's structured techniques are an improvement over earlier techniques. However, we can still make a mess with them. Most specifications for complex systems are full of ambiguities, inconsistencies, and omissions. More precise mathematically based techniques are evolving so that we can use the computer to help create specifications without these problems. As better, more automated techniques evolve, they need appropriate diagramming methods.

Just as advocates of different structured techniques tend to defend their method emotionally, so advocates of diagramming methods put up passionate defenses even when a different technique has superior qualities. Sometimes the advocates or owners of a particular diagramming technique defend it more like pagan priests defending a religion than like computer scientists seeking to advance their methods. It is very necessary to look objectively at the changes needed for full automation and integration of diagramming techniques, and to speak openly about the defects of earlier techniques.

Many of the diagramming techniques in common use are old and obsolete. The IBM diagramming template, which most analysts use, is two decades old. It contains symbols for ''magnetic drum,'' ''punched tape,'' and ''transmitted tape.'' It was created before data-base systems, visual display units, or structured techniques were in use. Most of the courses for systems analysts and programmers are obsolete.

That change is mandatory is certain because many of the diagramming techniques simply cannot draw some of the important constructs that we need to represent. We need an integrated set of diagramming standards with which we can express all of the constructs that are necessary for the automation of system design and programming.

STRUCTURED TECHNIQUES   The introduction of structured techniques into computing was a major step forward, referred to by its advocates as ''the structured revolution.'' The early structured techniques were pencil-and-paper methods. Today these techniques need automation. Designs should be created with the aid of a computer. The design should be of such a form that it leads to automated code generation. The *true* structured revolution is that which makes the programmer unnecessary by the use of code generators working from structured designs.

Structured techniques need structured diagramming. We must be able to draw all of the desirable constructs in a way that makes their meaning clear and obvious.

Most structured diagramming techniques support a top-down, structured development approach. They can describe a system or program at varying degrees of detail during each step of the decomposition process. They clarify the steps and the results of the decomposition process by providing a standardized way of describing procedural logic and data structures.

Structured diagramming techniques help developers deal with the large volume of detail generated during the program development process.

END-USER INVOLVEMENT

Particularly important in computing today is the involvement of end users. We want them to communicate well with systems analysts and to understand the diagrams which are drawn so that they can think about them and be involved in discussions about them.

Increasingly, some end users are likely to create their own applications with user-friendly fourth-generation languages. Where they do not build the application themselves we would like them to sketch their needs and work hand in hand with an analyst (perhaps from an information center), who builds the application for them. User-driven computing is a vitally important trend for enabling users to get their problems solved with computers [1].

For these reasons, diagramming techniques should be user-friendly. They should be designed to encourage user understanding, participation, and sketching. Many DP diagramming techniques have been designed for the DP professional only. To be user-friendly a diagram should be as obvious in meaning as possible. It should avoid symbols and mnemonics which the user may not understand.

PROGRAM DOCUMENTATION TOOLS

Structured diagramming techniques are very important as program documentation tools. They are used to define the program specifications and to represent the program design. They provide the blueprint for implementing the design into program code. Structure charts and Nassi–Shneiderman are two examples. Structure charts give the overview of the program structure; Nassi–Shneiderman diagrams draw the detailed internals. Together they describe the program organization structure and its internal workings. The programmer translates them into actual programming language instructions during the coding phase. Today it is desirable that we have *one* drawing technique that accomplishes *both* the overview diagramming *and* the diagramming of detailed internals. The overview diagram should be successively decomposed into the code structure. Action diagrams accomplish this.

Diagrams can give both high-level and detailed descriptions of a program.

For example, HIPO diagrams and structure charts can give a high-level overview of a program. They can be used to explain in general terms what major functions the program performs and what data and procedural components make up the program. On the other hand, pseudocode and Nassi–Shneiderman charts can give an instruction-level view of a program. They show where each program variable is initialized, tested, or referenced in the program code.

A high-level and/or a detailed view of a program are important, depending on the reader's purpose. If he is searching for a bug, detailed documentation may guide him to the exact location of the error. If he wants to determine in which of several programs a certain function is performed, high-level documentation may be the most helpful.

The tool used for high-level design is often different from the tool used for low-level design. For example, HIPO diagrams are suitable for high-level documentation, but at a low-level become too cluttered and do not show structured coding constructs. Nassi–Shneiderman diagrams show detailed program logic with structuring coding constructs but do not show overall program architecture.

However, to use different and incompatible techniques for high-level design and low-level design is generally undesirable. It dates back to an era when the program architect was a separate person from the detail coder. Today it is desirable that the high-level design be steadily decomposed into low-level design using the same diagramming technique. This decomposition should often be done by one person, preferably at a computer screen. As fourth-generation languages become more popular, the era of the separate coder will go. We need one diagramming technique that enables a person to sketch an overview of a program and decompose it into detailed logic. Action diagrams give this ability.

Data-base design has become very important and has spawned its own collection of diagramming techniques. The systems analysis process and the programming process need to be linked into the data-base design. Toward the end of the book (Chapters 21 and 22) we will see an integration of these. A data structure diagram is developed (often by a data administrator). A data navigation diagram is drawn on top of this. The data navigation diagram is converted (automatically) into an action diagram which is edited to produce executable code.

UTILITY OF DOCUMENTATION

Diagramming techniques produce both internal and external program documentation. *Internal documentation* is embedded in the program source code or generated at compile time. Program comments and cross-reference listings are examples of internal documentation.

External documentation is separate from the source code. HIPO diagrams and Warnier–Orr diagrams are examples of external program documentation. External program documentation, such as data flow diagrams and pseudocode, is often discarded once the program is developed. It is considered unnecessary and too expensive to keep up to date during the remainder of the system life cycle. If a program is well structured and properly documented internally, external program documentation becomes ignored.

Maintenance programmers mistrust most external documentation because they know that in practice it is seldom updated. Even the external documentation for a newly released system is unlikely to describe a program accurately.

Documentation can be trusted to be accurate in two cases. The first is when information about a program is automatically *generated from the code* (e.g., cross-reference listings, automatically generated structure charts, flowcharts, etc.). Keeping all the program documentation within or generated from the source code will make it more accessible and more accurate. The second is when the documentation is in the form of computerized diagrams or representations and the program is *generated* automatically from these. In one of these two ways (preferably the latter) the code and the diagrams are automatically linked and the diagrams *are* the documentation used for maintenance.

We should distinguish between different types of external documentation:

1. High-level structure versus detailed logic

2. Procedural structure versus data structure

High-level documentation will change very little, can be easily updated, and is a valuable source of information about a program throughout its life. Overview HIPO diagrams, Warnier–Orr diagrams, data flow diagrams, structure charts, data navigation charts, and action diagrams are valuable introductions to understanding a complex program.

Two types of high-level documentation are useful: control flow information and data structure information. Although often overlooked, data structure information can be the more useful of the two in aiding overall program understanding. For example, many common data processing application programs have a relatively simple control flow structure, but complex data structures. Experiments by Shneiderman support this position. He found that data structure documentation was more helpful to programmers than even more comprehensive control flow information (such as pseudocode) in overall program comprehension, including the procedural aspects [2].

The subject of data administration is particularly important. A data administrator is the custodian of an organization's data dictionary and data structures. He is responsible for maintaining and creating a model of data that will be used on many different projects. This central representation of data is the basic foundation stone of many programs. Having separate programs employ views of data extracted from a common data model ensures that data can be exchanged among these programs, and that data can be extracted from a common data base for management reporting and decision-support purposes. A clear diagramming technique is needed to represent the data models and views of data which are extracted from it.

The programmer must often navigate his way through a complex data base.

Data navigation diagrams, described in Chapter 21, give a diagramming technique for this, which is linked to the data model and to action diagrams which represent program structures that use the model.

If the data model is designed properly, the data structures should not usually change in disruptive ways throughout the system life. A canonical data model can be largely independent of individual applications of the data and also the software or hardware mechanisms which are employed in representing and using the data [3]. The data model and the data dictionary are valuable tools to aid program understanding. There may be many data navigation charts associated with one data model.

BATTLE WITH COMPLEXITY

Much of the future of computing is a battle with complexity. To push the frontiers forward, we have to learn how to build more complex systems. We cannot do this without harnessing the power of the computer itself.

The battle with complexity became more urgent in the world of hardware than in the world of software. Designers of the most complex chips and wafers could not succeed without highly sophisticated computer graphics tools. The 3-inch-square ceramic modules that hold chips in today's mainframes contain over a mile of wiring.

Whereas designers of hardware logic have taken very seriously their computer-aided tools, most analysts and programmers have not. Analysts and programmers often regard themselves as artists needing pencil-and-paper tools. We tolerate errors in their work to a much greater extent than we would in the work of hardware designers.

CHANGING SYSTEMS

Data processing systems are not static. They are modified and evolve with time like a living organism. We grow parts of them and prune other parts. We need to make constant adjustments.

This process of changing systems, however, has been very difficult with traditional methods. Changes tend to have unforeseen consequences. A change made in one part of a system causes errors in other parts. Often, the documentation is such that these consequential errors are neither foreseen nor detected until the system causes problems.

One of the most desirable features of computer-aided methodologies is that they should make systems easy to change. On-line diagrams make clear the structure of the data and the processes, and make it easy to modify these. When a modification is made, they automatically reveal where this has effects, and show what else has to be changed as a consequence of the modification.

FUNCTIONS OF STRUCTURED DIAGRAMS

Good diagramming techniques, then, have the following important functions:

- An aid to clear thinking
- Precise communication between members of the development team
- Standard interfaces between modules
- Systems documentation
- Enforcement of good structuring
- An aid to debugging
- An aid to changing systems (maintenance)
- Fast development (with computer-aided diagramming)
- Enforcing rigor in specifications (when linked to computerized specification tools)
- Automated checking (with computer-aided tools)
- Linkage to data administration tools
- Enabling end users to review the design
- Encouraging end users to sketch their needs clearly
- Linkage to automatic generation of code

A function that is achieved on only a few systems today ought to be emphasized and may become extremely important. The diagram, drawn on a computer screen, is decomposed into finer detail until executable program code can be generated from it *automatically*. Conventional programming then disappears.

This automation of programming has been achieved by HOS (higher-order software), whose objective is to generate bug-free code and create specifications which are internally bug free for exceedingly complex systems [4]. The automatic conversion of diagrams to code can clearly be improved. It has major advantages in speed of development, quality of code, and ease of maintenance.

REFERENCES

1. James Martin, *Application Development Without Programmers*, Prentice-Hall, Inc., Englewood Cliffs, NJ, 1982.

2. B. Shneiderman, "Control Flow and Data Structure Documentation: Two Experiments," *CACM*, Vol. 25, No. 1 (January 1982): 55–63.

3. James Martin, *Managing the Data-Base Environment*, Prentice-Hall, Inc., Englewood Cliffs, NJ, 1983.

4. James Martin, *Program Design Which Is Provably Correct*, Savant Technical Report 28, Savant Institute, Carnforth, Lancashire, UK, 1983.

2 TYPES OF DIAGRAM NEEDED

A systems analyst, like a carpenter, needs a number of different tools at his workbench. The tools that this book describes are among his most important— diagramming techniques that enable him to think clearly about complex system design.

Many systems analysts in the past have drawn one or perhaps two types of diagrams. In early training courses, flowcharts were the only type of diagramming taught. More recently, many analysts have learned only data flow diagrams and structure charts—both limited in what they can draw. We believe that a well-trained analyst should be comfortable in the use of the following techniques (or their equivalent):

- Decomposition diagrams
- Dependency diagrams
- Data flow diagrams
- Action diagrams (a replacement for structure charts)
- Data structure diagrams
- Entity-relationship diagrams
- Data navigation diagrams
- Decision trees and tables
- State-transition diagrams

This list does not include diagrams of machine configurations, data networks, or other representations of hardware.

EIGHT AREAS   Data processing requires techniques relating to the eight areas in Fig. 2.1. On the left is data; on the right is processing. At the top are high-level overviews. Reading down the chart, we arrive at levels of greater detail. Level 2 refers to structures which are independent of particular mechanisms, software, or program structures. These are sometimes referred to as *logical* structures: logical data models and logical process models. Other terms used are *conceptual* data models, software-independent data models, business flow models, and mechanism-independent or procedure-independent descriptions of processes.

The lower two layers relate to program-level structures. Here we refer to procedures rather than processes. It is necessary to draw the overall structure of programs (level 3) and the detailed logic and data usage (level 4).

The techniques for designing and representing data are discussed mainly in Chapters 19 and 20, although various specific methodologies such as Michael Jackson methodology (Chapter 12) have their own way of representing data. Chapter 20 discusses overview models of data. Chapter 19 discusses more detailed design of data.

The right-hand side of Fig. 2.1 needs a greater diversity of tools. At the top we are concerned with understanding how an enterprise works. We need a high-level view of corporate functions. Coming down to level 2 we need to be able to diagram business process, the dependencies among processes and the flow of data among processes. We do not yet design the mechanisms or programs.

Level 2 on both the data side and the process side is concerned with *logical* structures, not with physical implementation. There are multiple different ways to *implement* the logical structures. The diagrams at the left of level 2 need to reveal as clearly as possible the inherent logical structure of the data needed.

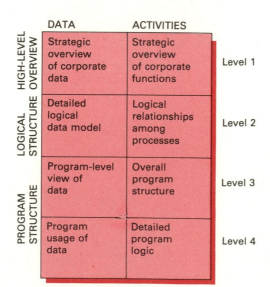

Figure 2.1 Areas to which diagramming techniques apply.

This representation of data should be fully normalized and as stable as possible [1]. The diagrams at the right of level 2 need to draw the relationships between the processes as clearly as possible. We need to understand what processes are needed to run the business, and how they interrelate.

Level 3 is concerned with the overall program structure. It is desirable to draw the structure of the program, making sure that it conforms to the best principles of structured techniques, before the detailed coding is done.

Level 4 is concerned with the detailed coding and logic of program. Here again it needs to be diagrammed before being coded. The right *type* of diagram makes it much easier to code. Indeed, with today's fourth-generation languages, the coding can be a quick, final step that can be performed in a computer-aided fashion, or in which the code is generated automatically.

TECHNIQUES FOR THE EIGHT AREAS   Figure 2.2 shows where the various techniques discussed in this book fit on the eight-area chart. Needless to say, any one corporation does not need all of the techniques listed in Fig. 2.2. It needs a carefully selected shortlist which covers all of the eight areas. The shortlist should be the basis of its training courses for systems analysts. The diagramming techniques need to fit together. A consistent set of techniques from Fig. 2.2 is needed which is as far as possible a common notation.

In doing our research for this book it became clear that many organizations had not made a good choice of diagramming techniques. The techniques they taught their analysts were difficult in various ways, and this did great harm to their efforts to use computers as effectively as possible.

Which techniques of those in Fig. 2.2 should an organization select? This is a complex question which the reader may decide when he has studied each technique in the book. Chapter 23 presents our own consumers' guide to the techniques. After analyzing the capabilities of the various methods, some stand out in mind more by what they *cannot* do than what they can. Some of the most publicized and popular techniques would be low on our consumers' choice list of preferences.

SUMMARY OF TYPES OF DIAGRAM   The following paragraphs summarize briefly the various techniques:

1. Decomposition Diagrams (Chapter 5)

High-level activities are decomposed into lower-level activities showing more detail. This top-down structuring makes complex organizations or processes easier to comprehend. Decomposition diagrams are a basic tool for structured analysis and design. Most decomposition diagrams are simple tree structures. It is useful

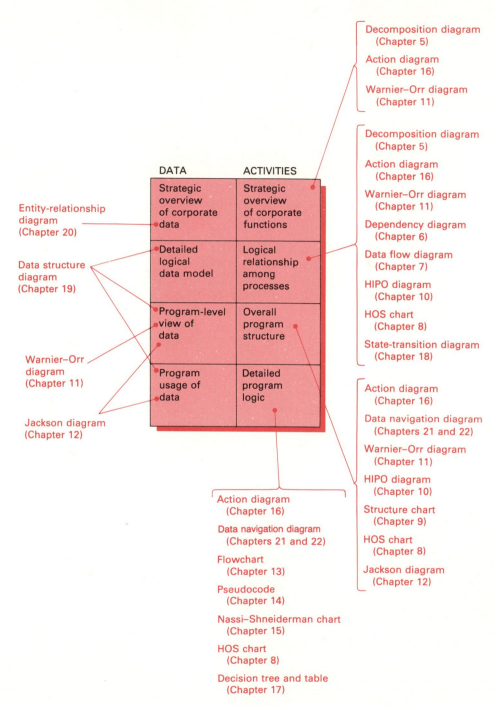

Decomposition diagram
(Chapter 5)

Action diagram
(Chapter 16)

Warnier–Orr diagram
(Chapter 11)

Decomposition diagram
(Chapter 5)

Action diagram
(Chapter 16)

Warnier–Orr diagram
(Chapter 11)

Dependency diagram
(Chapter 6)

Data flow diagram
(Chapter 7)

HIPO diagram
(Chapter 10)

HOS chart
(Chapter 8)

State-transition diagram
(Chapter 18)

Action diagram
(Chapter 16)

Data navigation diagram
(Chapters 21 and 22)

Warnier–Orr diagram
(Chapter 11)

HIPO diagram
(Chapter 10)

Structure chart
(Chapter 9)

HOS chart
(Chapter 8)

Jackson diagram
(Chapter 12)

Entity-relationship
diagram
(Chapter 20)

Data structure
diagram
(Chapter 19)

Warnier–Orr
diagram
(Chapter 11)

Jackson diagram
(Chapter 12)

DATA   ACTIVITIES

Strategic
overview
of corporate
data

Strategic
overview
of corporate
functions

Detailed
logical
data model

Logical
relationship
among
processes

Program-level
view of
data

Overall
program
structure

Program
usage of
data

Detailed
program
logic

Action diagram
(Chapter 16)

Data navigation diagram
(Chapters 21 and 22)

Flowchart
(Chapter 13)

Pseudocode
(Chapter 14)

Nassi–Shneiderman chart
(Chapter 15)

HOS chart
(Chapter 8)

Decision tree and table
(Chapter 17)

Figure 2.2   Areas in which different diagramming techniques are applicable.

to add other notation in some cases to show sequence, one-to-many decomposition, optionality, and conditions.

EXAMPLE: Fig. 24.1.

2. Dependency Diagrams (Chapter 6)

A dependency diagram has blocks showing activities and arrows between blocks showing that one activity is dependent on another. The arrows are often marked with data which are created by one activity and used by another (the reason for the dependency).

EXAMPLE: Fig. 24.2.

3. Data Flow Diagrams (Chapter 7)

This is a commonly used form of dependency diagram, which shows flows of data among processes or procedures. It shows data stores, and source and sinks of data.

EXAMPLE: Fig. 24.4.

4. Structure Charts (Chapter 9)

This is a hierarchical representation of a program structure. It forms part of the methodologies of Yourdon, and Gane and Sarson, where it is used in conjunction with a data flow diagram.

EXAMPLE: Fig. 9.2.

5. HIPO Diagrams (Chapter 10)

This diagramming technique uses a set of diagrams to show the input, output, and functions of a system or program. It consists of three types of diagrams.

- A *visual table of contents*, which is a tree-structured decomposition diagram.
- An *overview HIPO diagram*, which gives an overview of the input, process steps, and output.
- A *detail HIPO diagram*, which gives details of the input, process steps, and output.

EXAMPLE: Figs. 10.1, 10.2, and 10.3.

6. Warnier–Orr Diagrams (Chapter 11)

This diagram uses brackets to show the hierarchical decomposition of activities or data. This decomposition can represent a high-level overview of a program structure or detailed program logic. It forms the basis of the Warnier–Orr design methodology.

EXAMPLE: Fig. 11.5.

7. Michael Jackson Diagrams
(Chapter 12)

Jackson creates a tree-structured decomposition of the input and output data of
a program, and finds correspondences between these. The correspondences then
form the basis of another tree structure which becomes his program design.
Jackson methodology is one of several competing structured programming techniques.

EXAMPLE: Figs. 12.5 and 12.6.

8. Flowcharts (Chapter 13)

The flowchart was the predominant method of representing program logic prior
to the era of structured programming. Flowcharts are now in disfavor because
they tend to encourage GO TOs and nonstructured design. They have been replaced by structured English, Nassi–Shneiderman charts, and action diagrams.

For example, structure charts, Warnier–Orr diagrams, and Michael Jackson
diagrams give a top-down overview of program structure but not the detailed
logic. Flowcharts, structured English, and Nassi–Shneiderman charts give the
detailed logic but not the broad overview. Action diagrams give both the high
level and the detailed view.

EXAMPLE: Fig. 13.2.

9. Structured English and Pseudocode
(Chapter 14)

Diagramming techniques that cannot represent detailed program logic use structured English and pseudocode to show it.

EXAMPLE: Fig. 14.2.

10. Nassi–Shneiderman Charts
(Chapter 15)

This is a technique for drawing detailed program control structures. Unlike many
of the techniques for drawing overview program structures, Nassi–Shneiderman
charts can draw *condition*, *CASE*, *DO WHILE*, and *DO UNTIL* constructs.

EXAMPLE: Fig. 15.2.

11. Action Diagrams (Chapter 16)

This is a simple technique for drawing high-level decomposition, overview program structures, *and* detailed program control structures (*condition*, *CASE*, *DO
WHILE*, and *DO UNTIL* constructs). Most diagramming techniques cannot draw

both the overview structure of programs and the detailed control structures; action diagrams can.

EXAMPLE: Figs. 16.6 to 16.7, 16.13.

12. Decision Trees and Tables
(Chapter 17)

This is a technique for drawing logic that involves multiple choices or complex sets of conditions.

EXAMPLE: Fig. 24.13.

13. State-Transition Diagrams
(Chapter 18)

This is a technique for drawing complex logic that involves many possible transitions among states. Based on finite-state machine notation, neither decision trees nor state-transition diagrams are useful with every type of system or program; both relate to situations with certain types of complex logic.

EXAMPLE: Figs. 18.8 and 18.10.

14. HOS Charts

This is a controlled form of functional decomposition diagram (a tree structure) in which the data types which are the input and output of each block are shown and each decomposition is of a precise type defined with mathematical axioms. This is the basis of the HOS (higher-order-software) methodology with which specifications can be created for complex systems which are proven to have no inconsistencies or internal bugs. Bug-free code can be generated from these specifications.

EXAMPLE: Fig. 8.6.

15. Data Structure Diagrams
(Chapter 19)

These diagrams are designed to show detailed data structures. They are used for drawing normalized data and dependencies among data items. They can show derived data items.

EXAMPLE: Fig. 24.12.

16. Entity-Relationship Diagrams
(Chapter 20)

These diagrams draw entity types and associations among entity types (an entity is anything we store about data). Entity-relationship diagrams are the basis of high-level data models. Data structure diagrams show details of attributes in data models, data bases, or files.

EXAMPLE: Figs. 20.2 and 20.5.

17. Data Navigation Diagrams (Chapters 21 and 22)

These are diagrams showing the access paths through a data model or data-base structure which are used by a procedure. A first step in charting procedures which use data bases or multiple files. Appropriately drawn data navigation diagrams can be automatically converted to action diagrams.

EXAMPLE: Fig. 24.6.

STRUCTURED TECHNIQUES   Many of the diagramming techniques above relate to methodologies for structured analysis, design, and programming. Any data processing organization ought to be using structured techniques today. It is simply bad management not to. A choice has to be made about *which* structured methodology to make an installation standard.

It is important to understand that structured techniques themselves are changing. Many of the techniques introduced in the 1970s were deficient in serious ways and need to be supplanted with methodologies that:

- Are more complete
- Are faster to use
- Are based on sound data administration
- Are suitable for fourth-generation languages and application generators
- Enhance end-user communication
- Apply thorough verification techniques
- Solve the severe problems of maintenance
- And above all, are suitable for computer-aided design with interactive graphics.

The objective is to speed up the work of the analyst and programmer as much as possible, improve the quality of systems, and make systems easy to change. Figure 2.3 shows the evolution of structured techniques.

COMPUTER-AIDED DESIGN   The work of hardware logic designers is done today almost entirely with interactive computer graphics. The work of software logic designers has lagged behind, although computer graphics is equally relevant. Eventually, every systems analyst will be expected to use computer graphics tools.

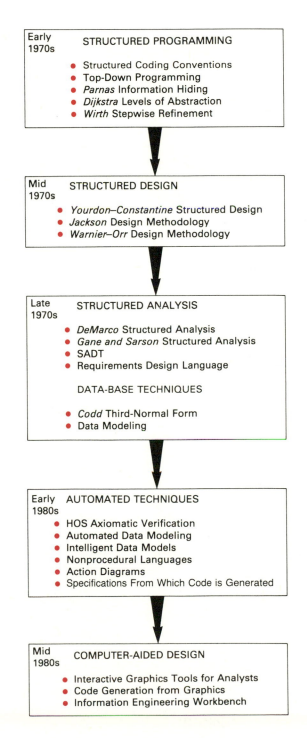

Figure 2.3 Evolution of structured techniques.

Some diagramming techniques that are in common use are of little value for computerized design. They are insufficiently precise or complete. Thorough automated validation cannot be applied to them. Their deficiencies have to be patched with techniques such as writing structured English or pseudocode. They do not form a basis for automatic code generation.

An analyst needs a family of tools like those listed on the first page of this chapter. With many of today's methodologies the tool kit is incomplete and the separate tools are incompatible. The analyst and designer use human intelligence to bridge the gaps between incompatible diagramming techniques. To benefit fully from computer-aided design, a complete and integrated set of diagramming conventions are needed.

COMPUTER GRAPHICS TOOLS

Computer graphics tools for analysts and system builders are destined to play a very important role. They vary greatly in their capabilities and value. We can categorize computer graphic tools as follows.

1. General-Purpose Drawing Tools

A tool with which static drawings of any type can be created.

EXAMPLE: LISADRAW, from Apple, Inc.

2. Tools for Using Dynamic Drawings

A *static drawing* is one without any built-in mechanisms. If one part of the drawing is changed or moved, it has no effect on any other part except perhaps to hide it or uncover it. A *dynamic drawing* is one with defined linkages between components or relationships between icons. When part of the drawing is changed or moved, other changes may occur *automatically*. A common example is that blocks on the drawing are connected by links (as with most of the diagrams in this book), and when one block is moved, the links automatically move with it such that they retain their logical meaning. Defined logical relationships among other elements of the drawing may also be automatically preserved. If the designer attempts to change the drawing in a way that makes it illogical or violates an integrity constraint, he will be automatically warned. The designer can build and manipulate dynamic drawings on a screen faster than static drawings.

EXAMPLE: EXCELERATOR from Index Technology, Inc.

3. Tools for Existing Methodologies

The preceding two categories are general-purpose tools that can be used with various design methodologies. The other categories in this list relate to tools for specific methodologies.

Several graphics tools have been built for implementing the methodologies that evolved in the 1970s: data flow diagrams, structure charts, Jackson diagrams, and so on. They provide the icons with which these diagrams can be created and modified quickly on the screen. Some of these tools employ menus, fill-in-the-blanks panels, or dialogues for helping create the diagrams, consider the integrity constraints, and generally do sound design.

EXAMPLE: STRADIS/DRAW from MCAUTO, which produces graphic support for Gane and Sarson's methodology using data flow diagrams and structure charts.

4. Computer-Enhanced Methodologies

There can be a world of difference between computerized versions of *hand* methodologies, and new methodologies designed to take advantage of the computer. The computer can employ large, precise libraries and can execute algorithms far too complex to be reasonable as manual methods. Computer graphics thus challenges us to redesign our methodologies.

Using a computer we can access large data models; automate the normalization and synthesis of these data models; extract and edit portions of the data models; create navigation paths through the data; generate screens, reports, and dialogues; generate and edit action diagrams; and apply mathematical axioms for enforcing correctness. We can automatically convert one type of diagram into another. High-level constructs can be automatically expanded into detail.

EXAMPLE: USE-IT from Higher Order Software Inc., which uses mathematical axioms to check the correctness of graphically built specifications and ensure that they are internally bug-free—a powerful technique that would be much too tedious to do by hand.

5. Graphic Designs that Generate Executable Code

Sufficient detail can be added to graphics diagrams to permit code generation. Several application generators use graphics to generate part of an application's code but not all of it: for example, reports, screens, dialogues, and data-base structures are generated from graphics. The program structure and logic can also be generated from dependency diagrams, data navigation diagrams, HOS charts, or action diagrams. The graphic tools may be linked directly to an interpreter or optimizing compiler, or may create code in a fourth-generation language which has its own interpreter or compiler.

6. An Integrated Family of Consistent Tools

It is desirable that the future system designer employ an integrated set of tools with which he can represent different aspects of system design. A consistent graphics notation is needed throughout this tool kit. The tool kit should be designed for automatic code generation. The tools should all run on the same computer with automatic conversion from one type of representation to another.

A DIAGRAMMATIC BASIS FOR AUTOMATION   To achieve CAD/CAM-like automation of system building, a consistent diagramming notation is needed for the various tools that represent different aspects of thinking about systems. The last chapter of this book recommends an integrated set of diagramming conventions which form a good basis for automation.

REFERENCE

1. James Martin, *Managing the Data-Base Environment*, Prentice-Hall, Inc., Englewood Cliffs, NJ, 1983.

3 FORMS OF STRUCTURED DIAGRAMS

INTRODUCTION The form or style of a diagram has a major effect on its usefulness. Complex charts can be drawn with a variety of techniques.

Human beings like to draw artistic charts with curvy arrows sweeping gracefully from one block to another. We position the blocks as our fancy takes us. Such charts may look nice but they cannot be easily maintained by computer. To enlist the computer's help in drawing and modifying our diagrams, the diagrams have to be reasonably disciplined.

Undisciplined diagramming becomes a mess when the diagrams grow in complexity (Fig. 3.1). Many diagrams that look nice in textbooks grow out of control in the real world when they have hundreds of blocks instead of 10. It is necessary to select forms of diagramming that can handle complexity.

FORMS OF TREE STRUCTURE A construct that appears in many different places in structured design is the hierarchy or tree structure. There are various different ways to draw a tree structure. The reader should be familiar with them and understand that they are equivalent.

A tree structure is used to indicate that an overall facility such as DOGMEAT CORPORATION includes lower-level facilities such as SALES DIVISION, MANUFACTURING DIVISION, and PLANNING DIVISION. One of these, such as MANUFACTURING DIVISION, includes still lower-level facilities, such as PURCHASING, PRODUCTION, and WAREHOUSE. We could draw this inclusion property as in Fig. 3.2.

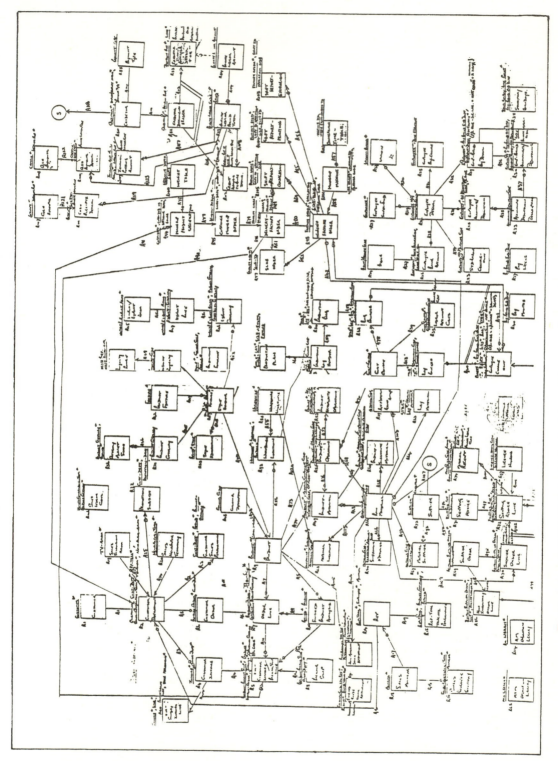

Figure 3.1 An example of a diagram that is an ill-structured mess. Large hand-drawn diagrams of excessive complexity tend to inhibit the making of necessary modifications.

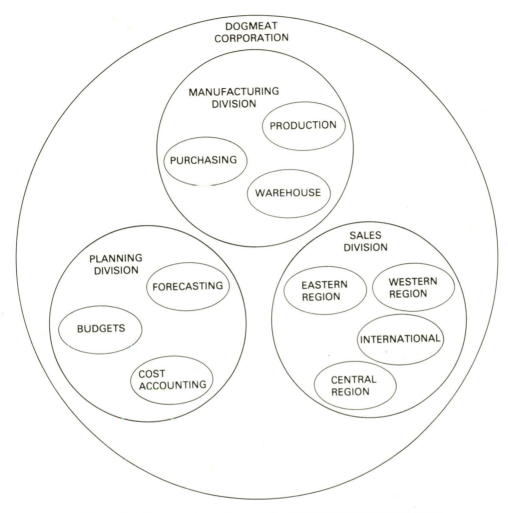

Figure 3.2   Tree structure showing what facilities include other facilities.

Figure 3.2 makes it clear what includes what, but would be clumsy to draw if there were many levels in the hierarchy. We can make it smaller by drawing it as subdivided rectangles, as in Fig. 3.3. This is neat and indicates clearly what includes what. This type of drawing is used in Nassi–Shneiderman charts in Chapter 15. A concern with Fig. 3.3 is that the writing of the lowest level is sideways. A normal printer cannot print sideways letters. It could print letters in

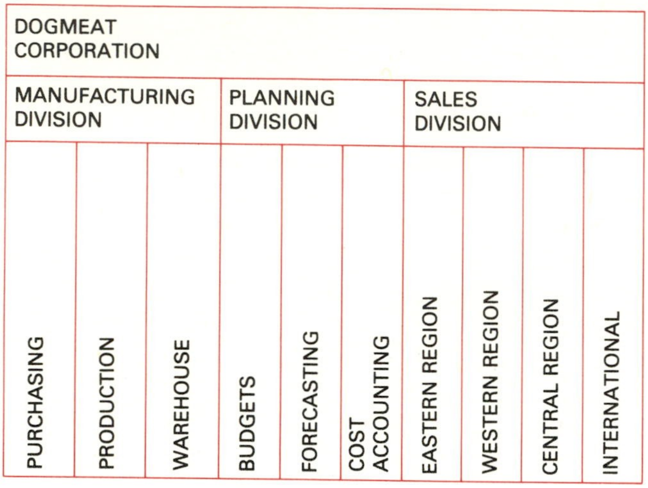

Figure 3.3   Box representation of a tree structure.

a column as in Japan:

P
U
R
C
H
A
S
I
N
G

But this becomes tedious to read.

A more usual way to draw a tree structure is as shown in Fig. 3.4. This form of diagram is used by many analysts and programmers. It is the basis of structure charts showing hierarchical structure in programs. It is also used to draw hierarchical data structures.

It looks good in textbooks because there it has a relatively small number of blocks. In real life there can be a large number of blocks at the lower levels and it will not fit on the width of a page. It might need paper six feet wide to draw it, and this is exactly what analysts, programmers, and data administrators do. They know that their work looks more impressive if it occupies a wall rather than resides in a binder. It is still more impressive if they draw it in multiple colors.

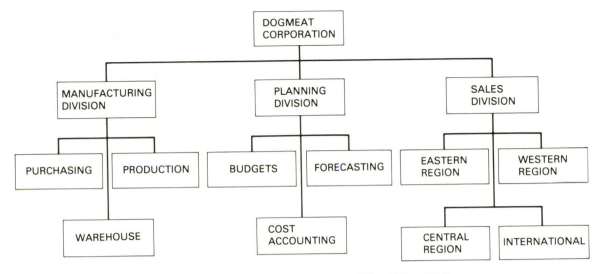

Figure 3.4　Common way of drawing the tree structure of Figs. 3.2 and 3.3.

**INHIBITION
OF CHANGE**

The problem with wall charts or hand-drawn works of art is that they are difficult to change. They inhibit change. In the design and analysis process there ought to be much change. The more a design can be discussed and modified in its early stages, the more the end result is likely to be satisfactory.

The best way to make a design easy to change is to draw it with a computer, on a screen, with software that makes it simple to modify. Automation of structured analysis, design, and programming using a workstation that draws graphics is generally desirable. We will need a printout of the diagrams created with a computer. With a CALCOMP plotter we can produce a six-foot wall chart, in color, that is sure to impress our colleagues.

Such plotters, however, are not freely available to every analyst and programmer. We would like to obtain a printout from our terminal or personal computer that does not produce six-foot wall charts unless we are prepared to stick many pieces of paper together with tape. In any case, wall charts are difficult to send to other people, or to take home.

**LEFT-TO-RIGHT
TREES**

We can solve the problem by turning the tree on its side. Figure 3.5 redraws the tree of Fig. 3.4. Now if there are many, many items at level three, it spreads out vertically rather than horizontally and can be printed with a cheap printer.

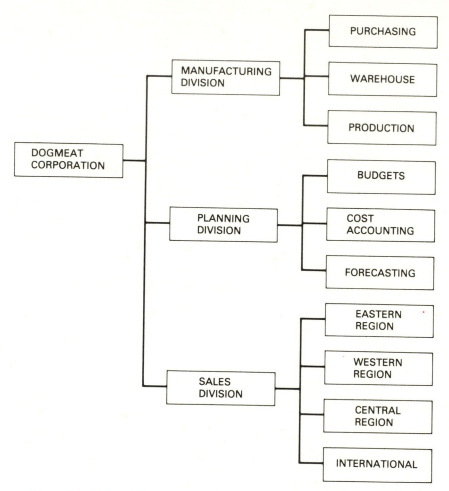

Figure 3.5   Left-to-right trees can be printed on normal-width paper even when they have many items at one level.

Tree structures are used for drawing organization charts of people. Here the tree may not be turned on its side because the person who runs the show wants to see his name at the top of the tree. With program or data structures there are no such ego problems, so left-to-right trees are fine.

Warnier–Orr notation, described in Chapter 11, draws a tree with brackets, as in Fig. 3.6. The left bracket implies that DOGMEAT CORPORATION *includes* MANUFACTURING DIVISION, PLANNING DIVISION, and SALES DIVISION, and so on.

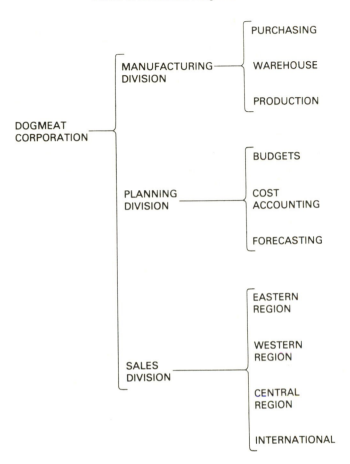

Figure 3.6   Warnier–Orr notation uses brackets to draw a tree structure, as in Fig. 3.5.

Suppose that there are many levels in the tree. The author was recently looking at one with 35 levels. A many-level tree drawn as in Figs. 3.4 or 3.6 would again spill off the right-hand edge of the page. To solve this we can have a more compact version of a left-to-right tree, as shown in Fig. 3.7. Incidentally, Fig. 3.7 solves the ego problem. The great leader can again be at the top of the tree.

The analyst has quite a number of lines to draw in Fig. 3.7. We would like him to have a small number of lines to draw so that he can make sketches quickly. Figure 3.8 shows a variant of Fig. 3.7 drawn with square brackets. This makes

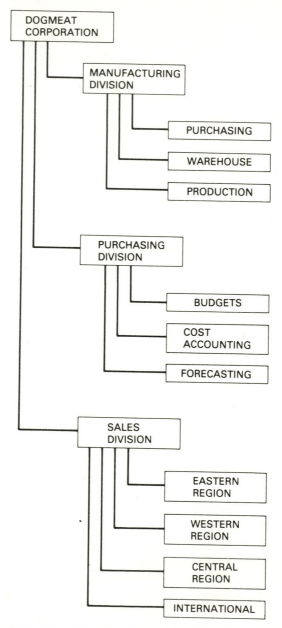

Figure 3.7 More compact version of Fig. 3.5. With this form, many levels could be included on one page.

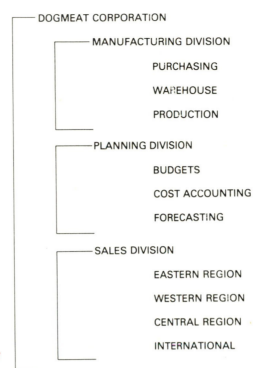

Figure 3.8   Square brackets enable us to draw
Fig. 3.7 with a small number of lines.

the hierarchical structure clear with a small number of lines. Each line of text in Fig. 3.8 could occupy much of the page. It could be a paragraph if so desired.

SEQUENCE OF
OPERATIONS

Structure charts (Chapter 9) are drawn as shown in Fig. 3.4. As commonly used, they do not show *sequence*. In much structured design it is necessary to show sequence: sequence of data items, sequence of programs modules, sequence of instructions. If blocks are clustered—more clustered than at the bottom right of Fig. 3.4—their sequence may not be clear unless precise sequencing rules are used. In Figs. 3.5 to 3.8 the sequence is clear. The items are implemented in a top-to-bottom sequence.

Figure 3.9 is much closer than the other diagrams to structured program code. If we remove the lines and boxes, it looks like Fig. 3.9.

We must eventually convert our diagrams into code. This conversion should change the format as little as possible to minimize the likelihood of making mistakes, and to make checking simple.

DOGMEAT CORPORATION

    MANUFACTURING DIVISION
        PURCHASING
        WAREHOUSE
        PRODUCTION
    PLANNING DIVISION
        BUDGETS
        COST ACCOUNTING
        FORECASTING
    SALES DIVISION
        EASTERN REGION
        WESTERN REGION
        CENTRAL REGION
        INTERNATIONAL

Figure 3.9 Figure 3.6 drawn without lines and boxes has structure like program code.

MESH-STRUCTURED DIAGRAMS

It is *easier* to make a mess with mesh-structured diagrams than with tree structures. Figure 3.10 shows a mesh structure, often called a network structure. In a tree structure there is one overall node, called the *root*, which we draw at the top in diagrams like Figs. 3.3, 3.4, 3.7, and 3.8, and at the left in diagrams like 3.5 to 3.8. This node has children, drawn below or to the right of their parent. They, in turn, have children, and so on, until the lowest or most detailed node is reached. This terminal node is called a leaf.

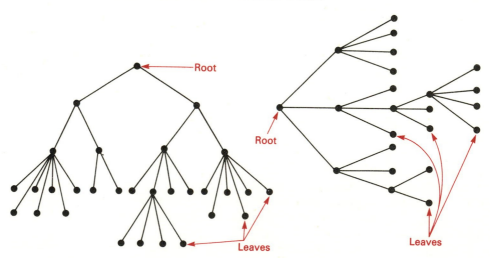

In a mesh structure there is no such neat ordering. A node may have multiple parents. Anything can point to anything. Chaos rules.

The objective of structured design is to prevent chaos ruling—to impose neatness on what otherwise might be disorder. GO TO instructions in programs can go to *anywhere*, permitting the programmer to weave a tangled mess. So structured design bans GO TOs and decomposes programs hierarchically.

Program structures, file structures, and document structures can, and should, be decomposed hierarchically. Unfortunately, there are some types of structures that cannot be decomposed hierarchically. A diagram showing how data flows through an organization cannot be beaten into a tree-like shape. A chart showing the relationship between entities in a data model is not tree-like.

Some structured methodologists almost refuse to accept the existence of anything that is not hierarchical. Because data models cannot be drawn with their hierarchical diagrams, these methodologists refuse to work with data models, and thus give little credence to the school of thought that regards data models as the foundation stone of modern DP methods.

COW CHARTS

Regardless of the difficulties, it is necessary to draw mesh-structured diagrams for some important aspects of design.

Figure 3.10 has 16 blocks. Its spaghetti-like structure makes it difficult to work with. Real life is worse and often such charts have several hundred items.

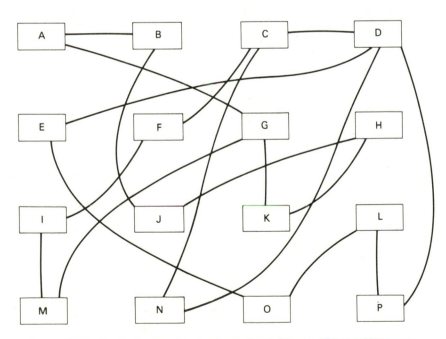

Figure 3.10   Mesh-structured (network structure) diagram. This would become a mess if it had a large number of nodes.

Such a chart may be cut up into pieces, but this does not clarify it unless it is done in a carefully structured way. The spaghetti-like pointers wander among the pages in a way that is confusing and difficult to follow.

Some designers create hand-drawn charts which are too big to redraw quickly and attempts to modify them create a rapidly worsening mess. The much modified chart is at last redrawn by hand and becomes regarded as a work of art—a triumphant achievement, but don't dare to modify it again!

The author has been horrified by some of the charts that data administrators keep. There is no question that these charts inhibit progress and improvement of the data structure. The data administrators will not dare to let end users propose changes to them. Sometimes these are called COW charts (can-of-worms charts). Most COW chart creators are impressed by their rococo masterpieces and pin them up on the wall.

NESTED CHARTS   To ease modification, certain types of charts can be nested. They are divided into modules that fit on normal-sized pages. In a tree structure, any of the blocks or brackets may be expanded in detail on another page. In a data flow diagram any of the processes may be expanded on another page.

Figure 3.11 shows a data flow diagram. Many data flow diagrams in practice have much larger numbers of processes (drawn as boxes in Fig. 3.11) and would require vast charts unless they were broken into nested pieces. Figure 3.12 shows Fig. 3.11 divided into two pieces.

The block labeled PROCESS 17 is expanded in a separate diagram, as shown. Blocks from this diagram, for example PROCESS 17.3, could be shown in further detail on another page. This subdividing of data flow diagrams is called *layering*. The layered structure of data flow diagrams within data flow diagrams within data flow diagrams is itself a hierarchy, like Fig. 3.2, and is often drawn as a tree structure.

Most analysts draw free-form data flow diagrams with curvy lines like those shown in Fig. 3.11. Using a computer to draw the diagrams speeds up the process and the software can perform some checking. Automated checking is very valuable in large projects with many data flow diagrams nested down to many layers. When the diagrams are computerized they are easy to change, which is important because of the expense and difficulty of maintenance.

Some computerized data flow diagrams need large paper and plotting machines to print them. They can, like other diagrams, be designed to spread out vertically rather than horizontally so that they can be printed by personal computers or normal printers, and can be put in three-ring binders rather than pinned on the wall.

Figure 3.13 shows a version of Fig. 3.12 designed for computer editing and normal printers.

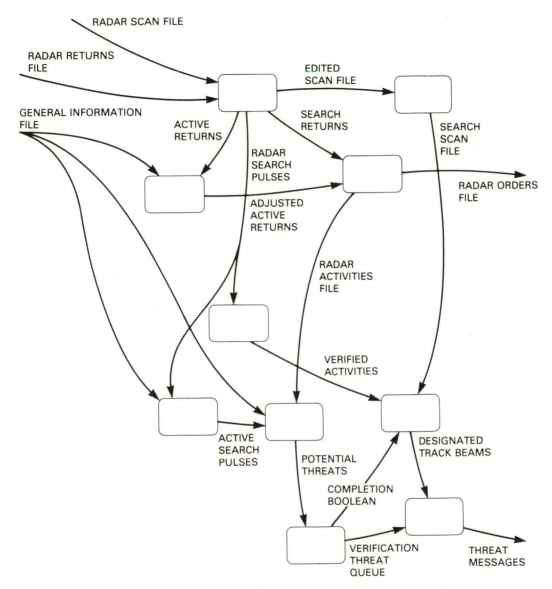

RADAR SCAN FILE

RADAR RETURNS
FILE

EDITED
SCAN FILE

GENERAL INFORMATION
FILE

ACTIVE
RETURNS

SEARCH
RETURNS

RADAR
SEARCH
PULSES

SEARCH
SCAN
FILE

RADAR ORDERS
FILE

ADJUSTED
ACTIVE
RETURNS

RADAR
ACTIVITIES
FILE

VERIFIED
ACTIVITIES

ACTIVE
SEARCH
PULSES

POTENTIAL
THREATS

DESIGNATED
TRACK BEAMS

COMPLETION
BOOLEAN

VERIFICATION
THREAT
QUEUE

THREAT
MESSAGES

Figure 3.11 Data flow diagram. The boxes are computer processes. The
arrows are data. It is desirable to divide complex mesh structures into modules
of less complexity. This diagram divided ("layered") in Fig. 3.12.

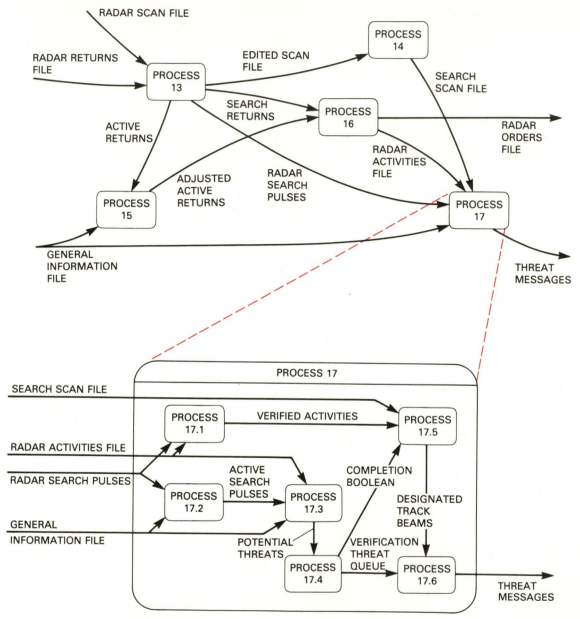

Figure 3.12 A complex diagram like this can be layered (i.e., divided in nested modules). The bottom diagram shows details of PROCESS 17 in the upper diagram.

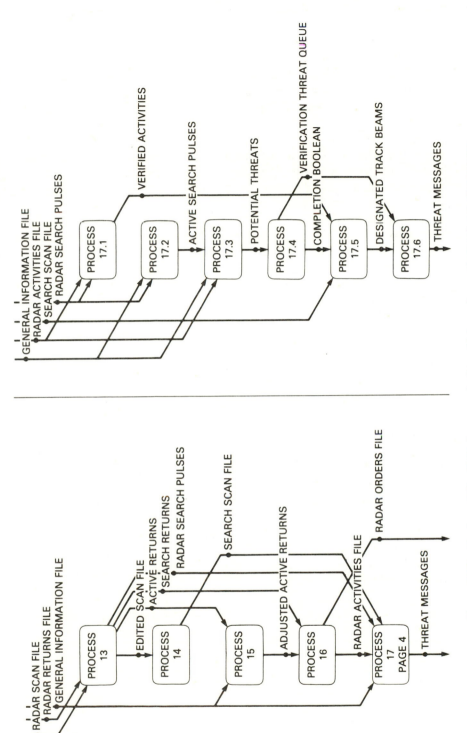

Figure 3.13 Version of Fig. 3.12 drawn vertically, which can be manipulated and changed by computer.

37

LATERAL VERSUS HIERARCHICAL SUBDIVISION

In designing systems we need to cope with diagrams that are too large and complex to be viewed at one time. The diagram has to be divided into pages or screens. This can be done in two ways, *laterally* and *hierarchically*.

In *lateral* subdivision, all pages show the same level. This is like a road atlas where every page contains a map of 5 miles to the inch. On a computer screen the user may *pan* across a large chart. (The movie director's words *pan* and *zoom* have come into our computer vocabulary.) The computer user may also skip among pages or "turn" pages.

Hierarchical subdivision is often more appropriate. A block on a high-level view can be expanded to show more detail. Figures 3.12 and 3.13 are illustrations of hierarchical subdivision. When a block is expanded into detail on a separate diagram, this diagram may show the shape of the parent and its title, as in Fig. 3.12.

Using computer graphics, the analyst may point to a block and ask to see it in more detail. Conversely, he may put a box around a group of blocks and shrink them to one block.

In Figs. 3.12 and 3.13 blocks are expanding such that the resulting diagram is of the same form. We will refer to this type of expansion as *exploding*. The converse process is called *imploding*. The menus with which an analyst manipulates graphics may contain the words EXPLODE and IMPLODE.

OTHER FORMS OF EXPANDING AND SHRINKING

The terms EXPLODE and IMPLODE refer to nesting in which the type of diagram remains the same. A block may also be expanded in other ways to show *different kinds* of detail. For this the words SHOW DETAIL, and the corresponding words HIDE DETAIL, are used, and may also be on the graphics menus. Similarly, the words ANNOTATE and DE-ANNOTATE may be used.

The words ZOOM IN and ZOOM OUT refer to expanding and shrinking a diagram within a window without changing its form, as with zoom lenses on cameras. We can thus have four different kinds of words on graphic menus for nesting and expanding diagrams.

ZOOM IN	ZOOM OUT	Expanding and shrinking without changing the diagram.
EXPLODE	IMPLODE	Changing to a more, or less, detailed diagram *of the same form*.
SHOW DETAIL	HIDE DETAIL	Changing to a more, or less, detailed diagram *of a different form*.
ANNOTATE	DE-ANNOTATE	Adding or removing descriptive matter.

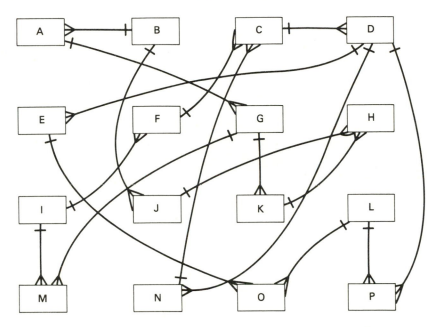

Figure 3.14   Data model chart drawn in an unstructured fashion.

DATA-MODEL CHARTS

While program structure diagrams and data flow diagrams can be modularized and nested, data models are more difficult to break into pieces. Teaching and insisting on modular design stopped programmers and analysts from papering the walls of their offices with vast, unruly charts, but now the data administrator is doing it.

Figure 3.14 shows 16 entities in a data model, and the associations among these entities. The line with a bar means a one-to-one association, for example:

$$\begin{array}{ll}
\text{TRANSACTION} & \hspace{-1em}\longmapsto\text{CUSTOMER} \\
\text{TRANSACTION} & \hspace{-1em}\longmapsto\text{SUPPLIER} \\
\text{EMPLOYEE} & \hspace{-1em}\longmapsto\text{EMPLOYER}
\end{array}$$

The line with a crow's-foot means a one-to-many association, for example:

$$\begin{array}{ll}
\text{CUSTOMER} & \hspace{-1em}\prec\text{TRANSACTION} \\
\text{SUPPLIER} & \hspace{-1em}\prec\text{TRANSACTION} \\
\text{SUPPLIER} & \hspace{-1em}\prec\text{CUSTOMER} \\
\text{CUSTOMER} & \hspace{-1em}\prec\text{SUPPLIER} \\
\text{EMPLOYER} & \hspace{-1em}\prec\text{EMPLOYEE}
\end{array}$$

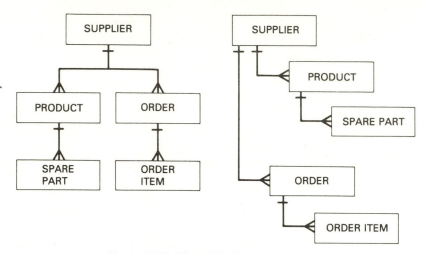

Figure 3.15   Hierarchical structures of data.

With this notation a hierarchical structure of data is drawn with the crow's-feet pointing down or to the right, and the one-to-many bars pointing up or to the left. Figure 3.15 shows this structure.

Such tree structures work well for representing a file or a document. One can draw a purchase order or bank statement with a tree structure, as we will see in later chapters when discussing Jackson and Warnier–Orr techniques.

However, it does not work by itself with data bases. The problem is apparent in Fig. 3.15. An order item is for a product. We would therefore like to draw a one-to-one link from ORDER ITEM to PRODUCT. We will do that, but then we no longer have a pure tree structure. Worse, a spare part *is* a product; it has a product number. We do not really want to regard it as a separate entity. Some PRODUCT records need to point to other PRODUCT records, indicating that the latter PRODUCT is a spare part for the former or that the former contains the latter. We can draw this by labeling two one-to-many links. Thus

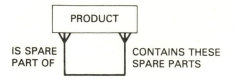

Figure 3.16 redraws the right-hand version of Fig. 3.15 to show these associations. Still more complications are introduced if we include customers in the diagram. The reader might like to explore this.

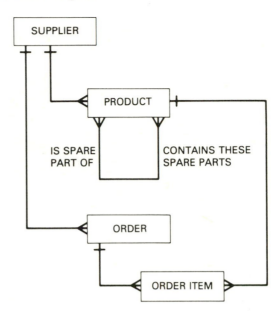

Figure 3.16 Figure 3.15 with associations between entities that are not hierarchical.

The data-base world, then, is full of constructs which cannot be drawn as pure tree structures. In drawing the associations among entity type network structures like Fig. 3.14 grow up. A medium-sized corporation has many hundreds of entity types and needs to represent them in a data model. The data model cannot be nested simply like a data flow diagram, so how should we draw a king-sized version of Fig. 3.14? How can we make it clear and more structured? An important variant of the same question is: How can we structure it so that it can be drawn and manipulated by a computer?

ROOT NODES

We can describe certain nodes in a mesh-structured chart as *root* nodes. A tree has one root node. We draw it at the top or left. A mesh structure has multiple root nodes. We can pull these also to the top or left. If we pull them to the top, we will end up with a diagram that spreads out horizontally, which is difficult to print. So let us pull them to the left.

The root of a tree structure is the only node with no one-to-one links leaving it. This can be seen in Fig. 3.15. We can use the same rule for discovering the root nodes of a mesh structure. Figure 3.17 marks the root nodes of Fig. 3.14.

We could remove the root nodes and their links from Fig. 3.17 and the remaining chart again identify the roots. We will call the original roots *depth 1* nodes, these second-level roots *depth 2* nodes, and so on.

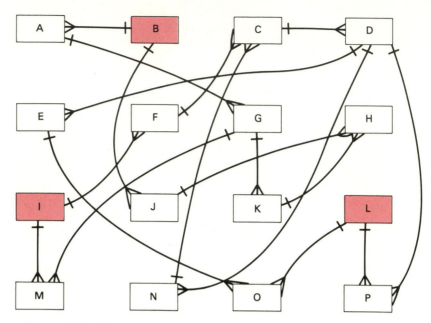

Figure 3.17   The root nodes of this mesh structure are shaded red. They are
the nodes that have no one-to-one links to another node.

A *depth 2* node can then be defined as a node that has a one-to-one link
pointing to a depth 1 node.

A *depth 3* node can be defined as a node that has a one-to-one link pointing
to a depth 2 node but no one-to-one link pointing to a depth 1 node.

A *depth N* node (N > 1) can be defined as a node with a one-to-one link
pointing to a depth (N − 1) node but no one-to-one link pointing to a lower-
depth node.

Figure 3.18 shows the depth numbers of the nodes on the chart in Fig. 3.17.

The depth 1 nodes are then plotted on the left-hand side of the chart. The
depth 2 nodes are offset by one offset distance. The depth N nodes are offset by
(N − 1) offset distances. The depth N node (N > 1) is plotted underneath the
depth (N − 1) node to which it points. The nodes under one depth 1 node form
a cluster. Links which span these clusters are drawn on the left of the chart, away
from the clusters, as shown in Fig. 3.19, which redraws Fig. 3.14.

The redrawing of a chart such as Fig. 3.14 begins with the identification
of the depth 1 nodes (no one-to-one links leaving them). Then the depth 2 nodes
can be marked; then the depth 3 nodes, and so on, until all of the nodes have
been given a depth number. The clusters under each root node are drawn, and
then the links spanning these clusters are added.

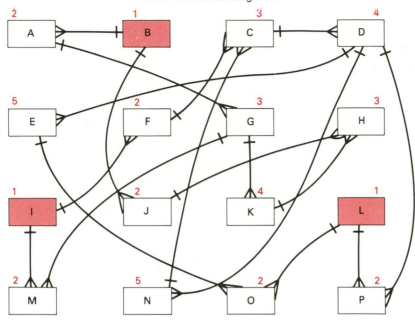

Figure 3.18 The red figures indicate the depth of each node, and are the basis for restructuring the diagram as shown in Figs. 3.19 and 3.20.

FIND THE TREES

Every mesh structure has little tree structures hidden in it. The process we have just performed might be called "Find the trees." In Fig. 3.14 you cannot see the trees for the forest. We can extract the trees, draw them as in Fig. 3.20, and then complete the diagram by drawing the links that span the trees.

In some mesh structures there is a choice of trees that could be extracted. A level 2 node might have two level 1 parents, for example. We make a choice based on which is the most natural grouping, or which is the most frequently traversed path. Usually, the trees extracted from a mesh structure turn out to be items that naturally belong together. Drawing an entity chart as in Fig. 3.20 clarifies these natural associations.

Figure 3.20, like Figs. 3.13 and 3.8, is a well-structured version of a chart that was messier. It is designed for computerized drawing and modification, and for a normal computer printer. Even though Fig. 3.20 looks relatively neat, an entity chart with hundreds of entities becomes complex, with too many lines to be easy to follow. Such a chart should be kept in a computer and the users should not normally see the whole chart. The computer presents them with a portion of the chart when they need it. Computerized extraction of subsets of large charts is extremely helpful.

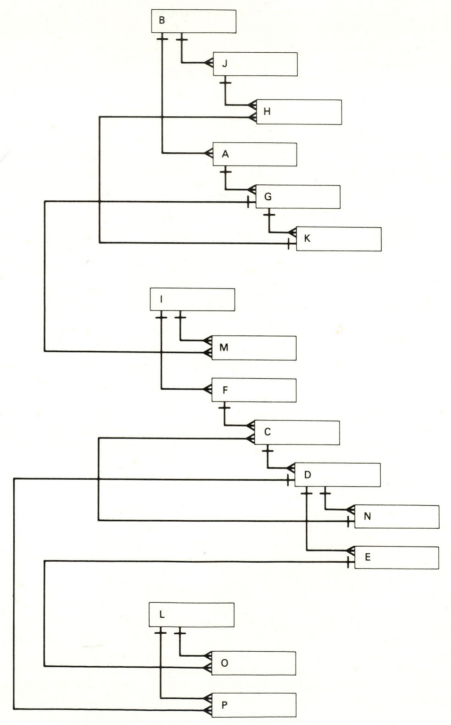

Figure 3.19 Figure 3.18 redrawn in a structured fashion.

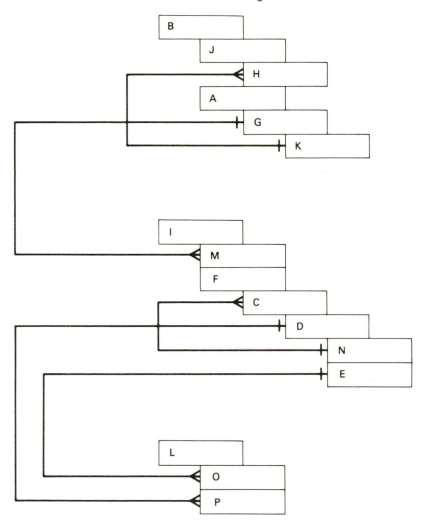

Figure 3.20   Figure 3.19 simplified by drawing its internal hierarchies without one-to-one links.

COMPUTER MAGIC   Once we use a computer all sorts of magic becomes possible in the manipulation of diagrams. A computer enables us to build big charts which it checks for consistency with a thoroughness far beyond that of most human beings. We can see an overview summary—just the highest layer. We can drop down to detail, descending through multiple layers. The computer can add color to show items of different meaning. It can highlight whole areas of a chart. It can extend the brackets of Fig. 3.8 into boxes. It can show detailed program logic or code within a box which we point to.

The diagramming technique may be designed so that human beings draw relatively few lines when they do it by hand, but once the computer goes to work it can dress up the diagrams to make them elegant and clear.

Some graphics software enables us to *zoom* into a diagram like a movie director with a zoom lens. As we move the cursor to an item and zoom in, the diagram changes to show us more detail—and more, and more, until perhaps we reach the coding level. Similarly, we can zoom out to see the overview. In some cases the diagram changes to another diagram of similar form; in other cases it changes to a diagram or display of a different form. The user may point at objects and expand them into windows showing detail.

When we change an item on a computerized chart we may have to change other items to keep the chart consistent. The software should point out all such consequential changes and insist that they are completed. In some cases it can make the consequential changes automatically.

The graphic symbols, sometimes called *icons*, can have logic associated with them so that when one change is made, consequential changes occur. If a box is moved, the lines and arrows connected to that box follow it. If a box or other symbol is connected illegally or in a way that raises questions, the designer is warned.

The designer is provided with advice, menus, guidelines, or other design aids. These may appear automatically or the designer may call them up when he wants them.

Once computer graphics is employed, the diagramming technique itself needs to be designed so that the computer can give the maximum help.

SYMBOLS WITH OBVIOUS MEANING   It is desirable that the symbols and constructs on a diagram have obvious meaning, as far as possible. For example, a diagram showing the components of a process needs to show *selection* and *repetition*. Some diagramming techniques do not show these. Some show them with symbols which are not obvious in meaning. With Michael Jackson diagramming, "*" means repetition and "o" means selection, when drawn in the top right-hand corner of a block. In Fig. 3.21, for example, one of the blocks marked "o" is selected. The block "PRINT BUZZ-WORDS," marked "*", is repeated multiple times.

Unless a key is written on the diagram, it would not be clear to an uninitiated reader what the "o" and "*" mean. People who once learned to read these charts forget the meaning of the "o" and "*" and would forget the meaning of other abstract symbols or mnemonics.

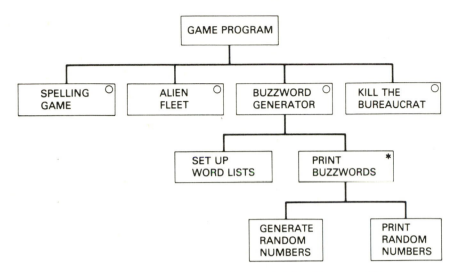

Figure 3.21 The "o" and "*" in the top right-hand corner of blocks on charts such as this do not have obvious meanings. The form of the diagram should be selected to make the meaning as obvious as possible to relatively uninitiated readers.

A map marker has the same problem. He chooses symbols which are as obvious in meaning as possible, such as

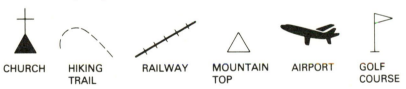

In addition, just to be sure, he puts a key on the diagram explaining the symbols.

A memorable means of illustrating repetition is to use a double line, double box, or double arrowhead. This is done in music. A double line in a score means repetition, thus:

In structured diagrams, a double block or double line at the head of a bracket could mean that the block or bracket is repeated.

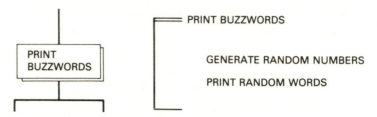

A memorable way of illustrating selection is to use a subdivided bracket:

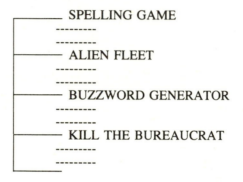

In the bracket above, *one* of the four items is performed, whereas in a nonsubdivided bracket everything is performed.

PRINT REPORT HEADER
PRINT REPORT BODY
PRINT TOTALS
PRINT STATISTICS

If a block or bracket is conditional, it is not enough to write a condition symbol or number (0,1) as on Warnier–Orr charts. The bracket should be able to show the nature of the condition:

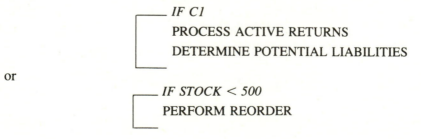

SUMMARY
Box 3.1 summarizes characteristics that are desirable in diagramming techniques.

Box 3.1 Properties of a Good Diagramming Techniques

- The diagrams are an aid to clear thinking.

- The diagrams can be manipulated easily on a computer graphics screen.

- End users can learn to read, critique, and draw the diagrams quickly, so that the diagrams form a good basis for communication between users and DP professionals.

- Hand-drawn diagrams are designed for speed of drawing; computer-drawn diagrams can have more lines and elaboration.

- The diagrams use constructs that are obvious in meaning, and avoid mnemonics and symbols that are not explained in the diagram.

- The diagrams can be printed on normal-sized paper. Wall charts of vast size are to be avoided because they tend to inhibit change and portability.

- Complex diagrams are structured so that they can be subdivided into easy-to-understand components.

- The overview diagram can be decomposed into detail; the designer does not necessarily have to resort to a different type of diagram to show the detail.

- The diagrams reflect the concepts of structured techniques.

- The diagrams are an aid to teaching computer methods.

- There is consistency of notation among all the different types of diagrams that an analyst needs.

- The diagrams should be a basis for computer-aided design and code generation.

4 A CONSISTENT DIAGRAMMING NOTATION

We have stressed that an enterprise should establish a set of standards for data processing diagrams. The standards should be the basis of the training given both to data processing professionals and to end users. Enterprise-wide standards are essential for communication among persons involved with computers, for establishing corporate or interdepartmental data models and procedures, and for managing the move into computer-aided design (CAD/CAP, CASA/CAP, or whatever is the favored acronym).

Many corporations have adopted diagramming conventions from methodologies of the past which today are inadequate because they are narrowly focused, ill-structured, unaware of data-base techniques, unaware of fourth-generation languages, too difficult to teach to end users, clumsy and time consuming, inadequate for automation, or, as is usually the case, tackle only part of problems that should be tackled.

This chapter summarizes the constructs that we need to be able to draw. Similar constructs are needed on many different types of diagram. A consistently drawn set of constructs can be used on the following basic tools:

- Decomposition diagrams (Chapter 5)
- Dependency diagrams (Chapter 6)
- Data flow diagrams (Chapter 8)
- Program structure diagrams (Chapter 10)
- Action diagrams (Chapter 16)
- Data structure diagrams (Chapter 19)
- Entity-relationship diagrams (Chapter 20)
- Data navigation diagrams (Chapter 21)

- Decision trees and tables (Chapter 17)
- State-transition diagrams and tables (Chapter 18)

This chapter discusses the constructs and how they are drawn. We will see these constructs appearing in many different types of diagrams in the subsequent chapters. Chapter 24 gives a recommended set of diagramming standards, and illustrations of diagrams drawn with the consistent standards.

STRUCTURED PROGRAM DESIGN

Structured programs are organized hierarchically. There is only one root module. Execution must begin with this root module. Any module can pass control to a module at the next lower level in the hierarchy—a parent module passes control to a child module. Program control enters a module at its entry point and must leave at its exit point. Control returns to the invoking (parent) module when the invoked (child) module completes execution.

A tree-structured diagram is used to draw the program modules that obey this orderly set of rules. As we saw in the preceding chapter, tree structures can be drawn in various ways. It is common to draw them as a set of blocks with the root block at the top, and each parent above its children. A neater way to show the flow of control is to draw them with brackets. Children are within, and to the right of, their parent bracket:

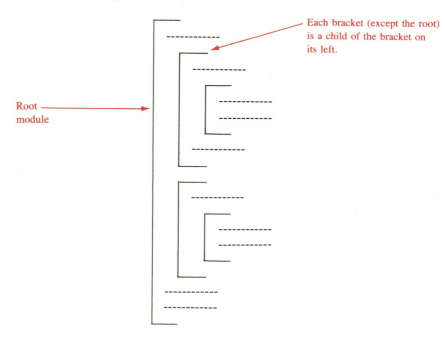

In creating structured programs four basic constructs are used.

SEQUENCE	Items are executed in the stated sequence.
CONDITION	A set of operations are executed only if a stated condition applies.
CASE	One of several alternative sets of operations are executed.
REPETITION	A set of operations is repeated, the repetition being terminated on the basis of a stated test. There are two types of repetition control, one (DO WHILE) where the termination test is applied *before* the set of operations is executed, the other (DO UNTIL) where the termination test is applied *after* the set of operations is executed.

Amazingly, some of the diagramming techniques used for representing structured programs cannot show these four basic constructs.

The four constructs can be shown very simply with brackets:

Sequence

    Everything in a
    bracket is executed
    in top-to-bottom sequence.

Condition

    IF X IS TRUE

    The contents of the
    bracket are executed
    if the condition at
    the head of the bracket
    applies.

Case

    IF A
    IF B
    IF C
    IF D

    One and only one
    of the divisions in
    this split bracket
    is executed.

Repetition

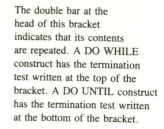

The double bar at the head of this bracket indicates that its contents are repeated. A DO WHILE construct has the termination test written at the top of the bracket. A DO UNTIL construct has the termination test written at the bottom of the bracket.

The words used in fourth-generation languages can be appended to the brackets. The diagram is thus edited until it becomes an executable program. Figure 24.9 shows an executable program drawn in this way.

This type of diagram is called an action diagram. At its initial stage it can be a tree structure representing a high level of overview or functional decomposition. It is successively extended until it becomes an executable program. This can be done in a computer-aided fashion, with software adding the words of a particular computer language. Action diagrams can be generated automatically from dependency diagrams, decomposition diagrams, data navigation diagrams, or decision trees.

BOXES

The family of diagramming tools discussed above use blocks to represent activities or data. To distinguish between activities and data, activities are drawn as round-cornered boxes, and data are drawn as square-cornered boxes.

ACTIVITY   DATA

ARROWS

Many types of diagrams have lines connecting boxes. Sometimes the lines have arrows on them, meaning flow or sequence. An arrow is used to mean *flow* or *sequence*. Flow usually implies that those activities are performed in sequence.

FLOW   SEQUENCE

CROW'S-FOOT

A crow's-foot connector from a line to a box is drawn like this:

It means that one or more than one instances of B can be associated with one instance of A. It is referred to as a *one-to-many association*. The term *cardinality* refers to how many of one item is associated with another. There can be one-to-one and one-to-many cardinality. Sometimes numbers may be used to place upper or lower limits on cardinality.

On diagrams of data, one-to-one cardinality is drawn with a small bar across the line:

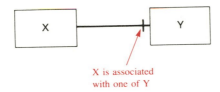

A line may have cardinality indicators in both directions:

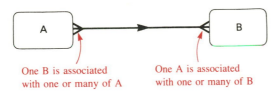

OPTIONALITY A dot in a gap at the start of a line means that that line is optional. The association may not exist.

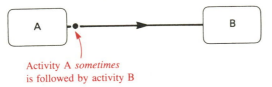

The reader may think of the dot as being an "o" for "optional."

When an arrow goes from activity block A to B, a circle against B means that B *sometimes* occurs without A:

The optionality circle is often placed in front of an activity because the activity can be triggered by many paths:

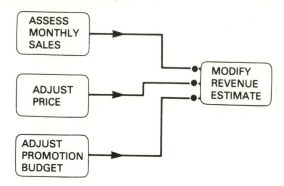

ZERO CARDINALITY

A data entity-type may be associated with *zero* of another data entity-type. For example, EMPLOYEE may be associated with one or *zero* WIFE. He may have *zero*, one, or many children. This is cardinality information; it is shown by a "0" on the link as part of the cardinality symbol:

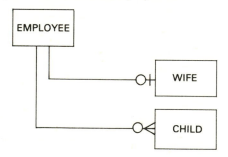

The "0" means the same, in effect, as an optionality dot on a link "—●——." Data model designers prefer to put all the cardinality information in one place and so draw ——○⊢[and ——○⋖[rather than optionality dots.

THINK OF LINKS AS SENTENCES

When drawing a link between blocks the designer should think of a sentence which describes the link. The positioning of the symbols we have described relates to such sentences. Thus:

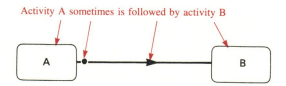

Activity A sometimes is followed by activity B

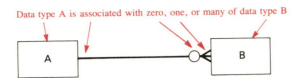

Sometimes a verb is written on the link to state its meaning:

Occasionally, there are two links between blocks, which need to be distinguished with labels:

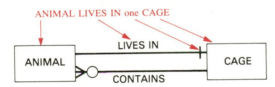

There are often links in opposite directions between blocks:

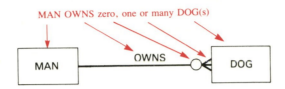

These are usually combined into one line:

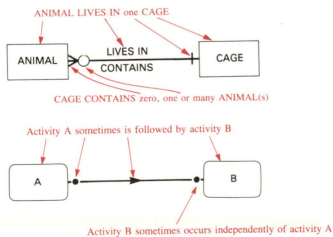

On horizontal lines the upper label is read when traversing *from left to right*; the lower label is read when traversing *from right to left*.

On vertical lines the left-hand label is read when traversing *down* the line; the right-hand label is read when traversing *up* the line.

**MUTUAL
EXCLUSIVITY**

Sometimes a block is associated with one of a group of blocks. This is indicated with a branching line with a circular dot at the branch.

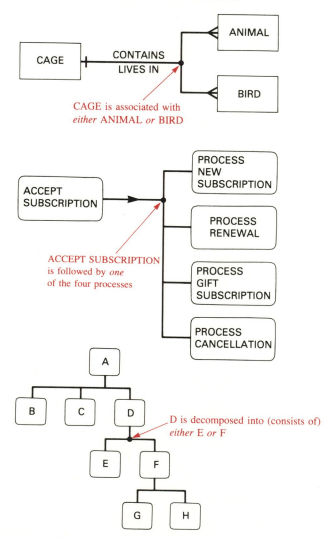

The reader may think of the dot as being a small "o" for "or."

**MUTUAL
INCLUSIVITY**

A branch without a dot means that *all* of the items branched to are required:

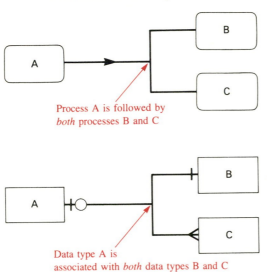

Various combinations of mutual exclusivity and inclusivity are possible, as shown in Fig. 4.1.

CONDITIONS

Condition statements are often associated with the optionality dots and mutual exclusivity dots:

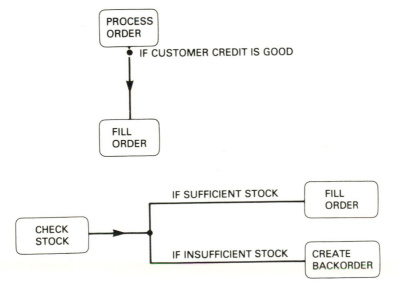

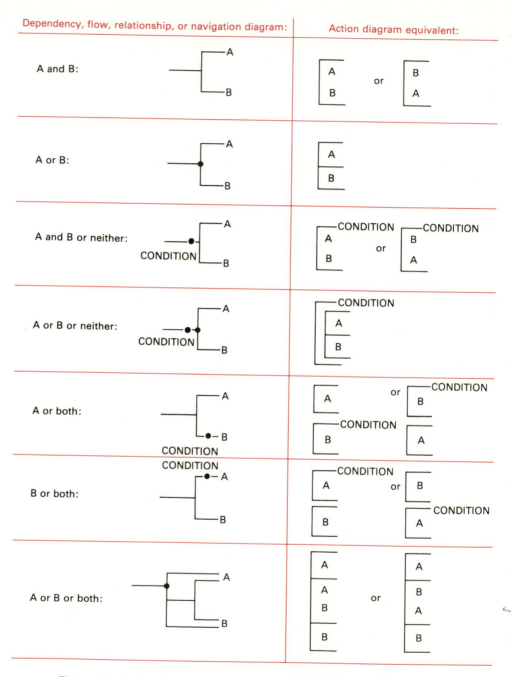

Figure 4.1   In the choice of two lines A or B, a number of possibilities exist.

When we convert dependency diagrams or data navigation diagrams into action diagrams ready for creating executable code, the condition statement will appear on the action diagrams.

The case structure of action diagrams is a similar shape to the branching mutual exclusivity line:

FLOW OR DEPENDENCY DIAGRAM

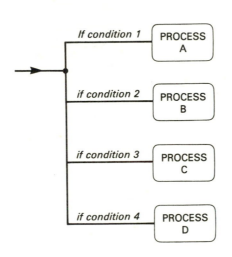

EQUIVALENT ACTION DIAGRAM

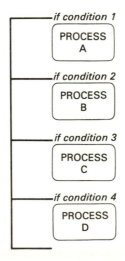

SEQUENCE

One of our objectives is to convert the diagrams we create into executable code as automatically as possible. Where we have diagrams showing activities, such as dependency diagrams, data flow diagrams, or data navigation charts, we need to know the *sequence* in which the blocks are executed. Where there is a single arrow from one block to another, the sequence is clear. Where there are multiple lines leaving a block, or where a line has a branch in it, the sequence is not clear:

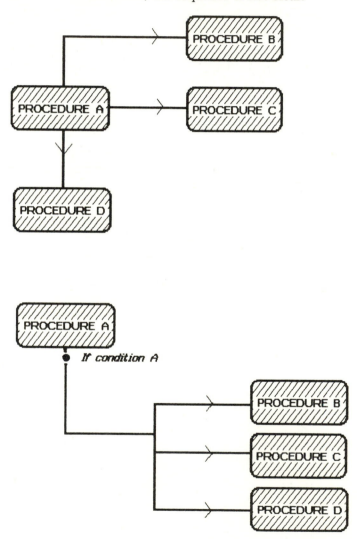

In order to make the sequence of execution clear, numbers may be attached to the line, thus:

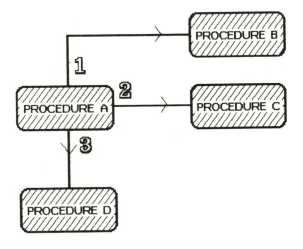

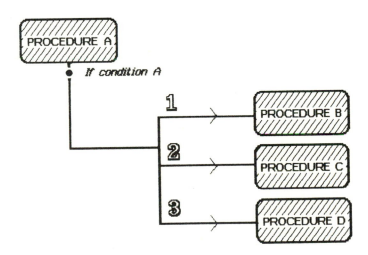

Later in the book we will see these numbers on dependency diagrams and data navigation charts.

CONVERSION OF DIAGRAMS TO EXECUTABLE CODE

There is a correspondence among the differing diagramming types. They need to be associated in order to automate as fully as possible the tasks of the analyst and programmer. It is this drive toward computer-aided design that makes it so important to have consistent notation among the different types of diagrams.

A data navigation diagram (Chapter 21) is drawn using an entity-relationship diagram (Chapter 20). A data navigation diagram can be converted automatically into an action diagram (Chapter 16). Similarly, a dependency diagram (Chapter 6) can be converted automatically into an action diagram. An action diagram is edited in a computer-aided fashion until it becomes executable code. This computer-aided progression from high-level overview diagrams or data administrator's data models to executable code is what makes it possible to increase the productivity of the systems analyst by such a large amount.

Later chapters contain examples of the automated conversion to action diagrams and then to executable code.

SUMMARY

In this chapter we have described the elements of a consistent diagramming notation. These elements reappear in many of the following chapters. Some diagramming notations are different from this consistent notation. These include HIPO diagrams, Warnier–Orr diagrams, Michael Jackson diagrams, and Nassi–Shneiderman charts. It was tempting to modify the style of some of these diagrams to make them conform to our consistent notation, but we have not done so. We have presented them as they are normally used.

Figure 24.14 shows a template for use in data processing diagramming.

5 DECOMPOSITION DIAGRAMS

One of the simplest of analysts' diagrams is the decomposition diagram. A high-level organization, function, or activity is decomposed into lower-level organizations, functions, or activities. The lower we go in this hierarchy, the greater the detail revealed. Tree structures are used to show the decomposition and can be drawn in the various ways described in Chapter 3.

In decomposition diagrams a parent block *is composed of* its offspring blocks. It could be described as a "composed-of" diagram. The offspring together completely describe the parent. In some other tree structures this is not true. In *some* program structure diagrams a parent block *invokes* its child blocks but may itself contain functions that are not in the child blocks; the child blocks are, in effect, subroutines.

Most structured design employs a form of decomposition. A high-level representation of an activity is decomposed into lower-level, more detailed activities; these are decomposed further; and so on. The term "decomposition" is sometimes used with an adjective which says what is being decomposed: for example, "functional decomposition," "process decomposition," "procedure decomposition," and "data decomposition."

Decomposition diagrams are used to show organization structures, system structures, program structures, file structures, and report structures. Similar diagrams are sometimes drawn for the decomposition of data and the decomposition of processes. The diagram in the front of this book (pages vi–vii), showing the book structure, is a decomposition tree.

FUNCTIONS, PROCESSES, AND PROCEDURES

Decomposing the activities of an enterprise may start at a high level. A diagram illustrates the basic *functions* of the organization. Each function may be further decomposed into the *processes* that are necessary to accomplish that function. The diagram at this level shows the processes, but does not show the detailed procedures or mechanisms by which they will be accomplished. Going lower in a hierarchy, we break the processes into *procedures* and decompose computerized procedures into program structures. We use the word "activity"as a generic term meaning "function," "process," or "procedure."

Functions

Functions refer to major areas of activity in a corporation, such as engineering, production, research, and distribution. One medium-sized manufacturing company lists its functions as follows:

- Business planning
- Finance
- Product planning
- Materials
- Production planning
- Production
- Sales
- Distribution
- Accounting
- Personnel

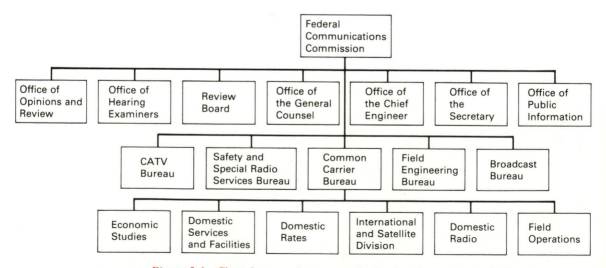

Figure 5.1 Chart decomposing an organization into its component bodies.

For clear distinction between the words *function* and *process*, *function* may be referred to as *functional area*. A simple tree-structured diagram is normally used to show the functional areas in an enterprise.

An organization chart decomposes an organization into its component bodies, as shown in Fig. 5.1. The chart of functional areas is usually different from the organization chart. It represents the basic functions that have to be carried out, regardless of what organization groups or departments exist. Individual units or divisions, such as *CATV Bureau* in Fig. 5.1, may be subdivided into their functional areas.

Processes

A process is a specific activity that has to be carried out. In each functional area there are many processes. Figure 5.2 is a tree-structured diagram showing organizations, functional areas, and processes.

A diagram such as Fig. 5.2 is sometimes referred to as a *corporate model* or *enterprise model*. It shows the functions and processes that are necessary in running the enterprise, without showing any detail about how they are accomplished.

Enterprise models differ in the degree of detail which they represent. Some show not only the functional areas and processes, but also lower levels in the tree, indicating detailed activities that are carried out in each process. Some information-planning methodologies refer only to functional areas and processes; some refer to functional areas, processes, and activities. The methods that give the best results are in general the more detailed ones. These need computerization; there is too much detail to manipulate by hand.

Figure 5.2 lists 42 processes. A large complex corporation might have 30 or so functional areas and 150 to 300 processes. The chart needs to be drawn as a tree which progresses *across* the page, like Fig. 5.2, rather than as a *vertical* tree which progresses *down* the page like Fig. 5.1.

Identification of functional areas and processes should be independent of the current organization chart. The organization may change but still have to carry out the same functions and processes. Some corporations reorganize traumatically every two years or so. The identification of functions and processes should represent fundamental concern for how the corporation operates, independently of its current organization chart (which is often misleading). In some organizations there needs to be basic questioning about whether the functions and processes perceived are sound.

The names given to the processes should be action-oriented nouns such as those in Fig. 5.2. They often end in the suffixes *-ing*, *-ion*, or *-ment*.

Sub-processes

In each business process, a number of activities take place. One of the processes in Fig. 5.2 is purchasing, for example. In this process, activities such as the

ORGANIZATION	FUNCTIONAL AREAS	PROCESSES
CATFOOD DIVISION	BUSINESS PLANNING	Market analysis Product range review Sales forecasting
	FINANCE	Financial planning Capital acquisition Funds management
	PRODUCT PLANNING	Product design Product pricing Product specification maintenance
	MATERIALS	Ingredient planning Purchasing Receiving Inventory control Quality control
	PRODUCTION PLANNING	Capacity planning Plant scheduling Work flow layout
	PRODUCTION	Ingredients control Mixing and cooking Canning Machine operations
	RESEARCH	Psychological testing Narcotics Package design Cat farm
	SALES	Territory management Selling Sales administration Customer relations
	DISTRIBUTION	Finished stock control Order servicing Packing Shipping
	ACCOUNTING	Creditors and debtors Cash flow Payroll Cost accounting Budget planning Profitability analysis
	PERSONNEL	Personnel planning Recruiting Compensation policy

Figure 5.2 Organization, functional areas, and processes. The processes may be further subdivided, as shown in Fig. 5.3.

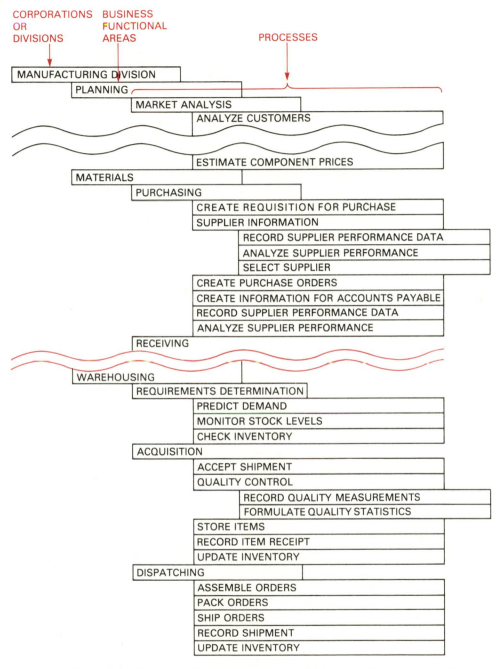

CORPORATIONS BUSINESS
OR FUNCTIONAL
DIVISIONS AREAS PROCESSES

MANUFACTURING DIVISION
  PLANNING
    MARKET ANALYSIS
      ANALYZE CUSTOMERS

      ESTIMATE COMPONENT PRICES
  MATERIALS
    PURCHASING
      CREATE REQUISITION FOR PURCHASE
      SUPPLIER INFORMATION
        RECORD SUPPLIER PERFORMANCE DATA
        ANALYZE SUPPLIER PERFORMANCE
        SELECT SUPPLIER
      CREATE PURCHASE ORDERS
      CREATE INFORMATION FOR ACCOUNTS PAYABLE
      RECORD SUPPLIER PERFORMANCE DATA
      ANALYZE SUPPLIER PERFORMANCE
    RECEIVING

  WAREHOUSING
    REQUIREMENTS DETERMINATION
      PREDICT DEMAND
      MONITOR STOCK LEVELS
      CHECK INVENTORY
    ACQUISITION
      ACCEPT SHIPMENT
      QUALITY CONTROL
        RECORD QUALITY MEASUREMENTS
        FORMULATE QUALITY STATISTICS
      STORE ITEMS
      RECORD ITEM RECEIPT
      UPDATE INVENTORY
    DISPATCHING
      ASSEMBLE ORDERS
      PACK ORDERS
      SHIP ORDERS
      RECORD SHIPMENT
      UPDATE INVENTORY

Figure 5.3 Enterprise chart: corporations or divisions, functional areas, and processes. The lowest-level process is sometimes called an elementary process. Its name should begin with a verb. A procedure (not necessarily computerized) is designed for each elementary process.

following occur:

- Create requisitions for purchase
- Select suppliers
- Create purchase orders
- Follow up the delivery of items on purchase orders
- Process exceptions
- Prepare information for accounts payable
- Record supplier performance data
- Analyze supplier performance

There are typically 5 to 30 subprocesses for each business process in Fig. 5.2. There may be several hundred activities in a small corporation and several thousand in a large, complex one. Figure 5.3 illustrates this hierarchical breakdown of processes. The lowest level process may be called an *elementary* process.

In many corporations the activities have never been charted. When they are listed, and related to the data they use, it is usually clear that much duplication exists. Each area of a corporation tends to expand its activities without knowledge of similar activities taking place in other areas. Each department tends to create its own paperwork. This does not matter much if the paperwork is processed manually. However, if it is processed by computer, the proliferation of separately designed paperwork is harmful because it greatly increases the cost of programming and maintenance. A computerized corporation ought to have different procedures from a corporation with manual paperwork. Most of the procedures should be on-line with data of controlled redundancy and minimum diversity of application programs. The entry of data in a terminal replaces the need to create multiple carbon copies of forms that flow among locations. Information becomes instantly available and procedures should be changed to take advantage of this.

A chart like Fig. 5.3 becomes large. It will be changed and added to many times and so needs to be kept in computerized form. From the overall computerized chart, small charts can be extracted to show the corporation and its functional areas, an overview of one functional area, or a detailed breakdown of one process.

End users and user management relate well to charts such as Figs. 5.1, 5.2, and 5.3. They can be encouraged to draw such charts and, where useful, decompose them into the detail necessary for planning computer programs.

Procedures

The term *process* refers to an activity without describing the mechanisms by which it is accomplished. The term *procedure* refers to a specific method of accomplishing the process; in other words, it refers to the design carried out by a systems analyst. The procedure may refer to documents, data flow, screen

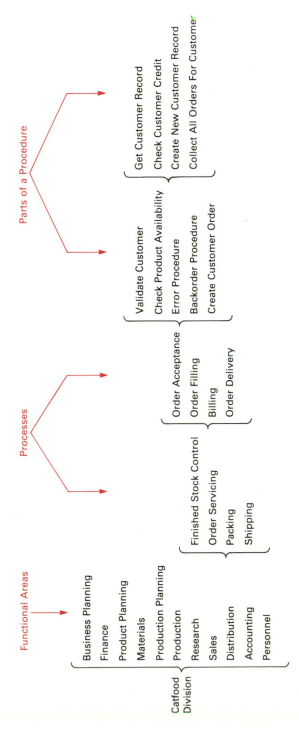

Figure 5.4 Decomposition progressing from the highest level to the level of program modules. This type of diagram shows insufficient detail at the procedure level.

interaction, and program steps. The procedure will be decomposed until a program structure is obtained. Not all procedures are computerized. A process refers to *what* must be accomplished.

TOP-DOWN DESIGN

Analysis and design can start at the highest functional areas in an enterprise and proceed into finer levels of detail until it represents program modules, and then show the detailed structure of those program modules. In many cases, top-level planning is not done, and decomposition starts at a lower level, sometimes within one department. It is desirable to accomplish top-level planning of data in an organization—the data administration function. The strategic planning of data ought to link the decomposition of functions and processes [1].

Figure 5.4 shows this. The functions at the left are those in Fig. 5.2. The process ORDER SERVICING is decomposed into lower-level processes, and one of these is decomposed into components of a procedure. Figure 5.5 shows this procedure drawn as a hierarchical "structure chart" of the type that many programmers use.

Box 5.1 summarizes the definitions of functional areas, processes, and procedures. To avoid confusion among these, they may be differentiated in diagrams by using different shading of the blocks, as in Box 5.1. The shadings correspond to the strategic level (no shading), analysis level (dotted shading),

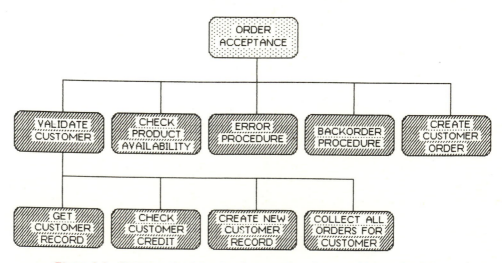

Figure 5.5   Structure chart showing the procedure decomposition on the right of Fig. 5.4. More detail is needed for effective diagramming of procedures.

BOX 5.1   Definition of functional area, process, and procedure

Activity is a generic term meaning procedures, processes, or overview operations at the functional area level.

Functional Area

A group of activities and decisions which together completely support one aspect of running the enterprise.

Process

An activity, uniquely definable, which can be accomplished completely in a finite time. The process does not indicate the precise method by which the results are accomplished.

Procedure

A method by which a specific process may be carried out.

Recommended Diagramming Conventions:

Functional area:

```
DISTRIBUTION
```

Process:

```
ORDER
ACCEPTANCE
```

Procedure:

```
CREATE NEW
CUSTOMER
RECORD
```

and the design level (crosshatching), as discussed in Chapter 4. Such shading can be added to diagrams automatically when a computer graphics tool is used.

ADDITIONAL CONSTRUCTS ON DECOMPOSITION DIAGRAMS

The simple decomposition of Figs. 5.1 to 5.5 is good enough for creating an overview of an enterprise's functions and processes, but it is not good enough for diagramming the details of procedures. Other types of diagrams are used for showing procedures. Decomposition diagrams can be made more generally useful by including additional constructs in them.

Optionality

A certain branch of a tree may be optional. It may or may not exist. We show this with a dot on the branch (''o'' for optional)'':

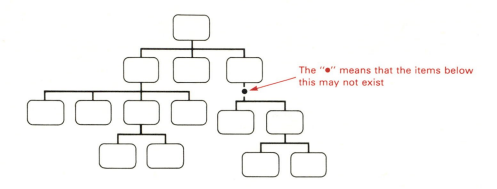

The "●" means that the items below this may not exist

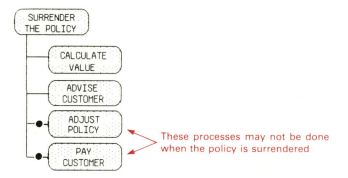

These processes may not be done when the policy is surrendered

Conditions

A condition may be written against the optionality dot to show when certain blocks are omitted:

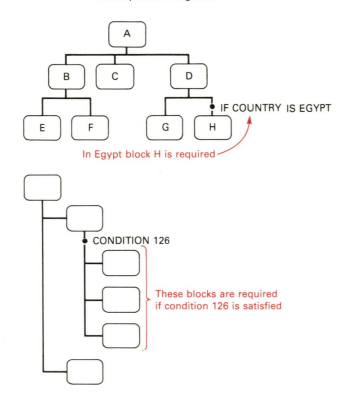

In Egypt block H is required

IF COUNTRY IS EGYPT

CONDITION 126

These blocks are required
if condition 126 is satisfied

Mutual Exclusivity

Sometimes *either* one item *or* another are permissible, but not both; or one item from a group is permissible. This is indicated with a dot on a branching line:

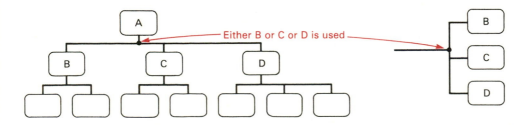

Either B or C or D is used

A branch without a dot means mutual *inclusivity*; that is, if one of the items branched to is used, the others are also used.

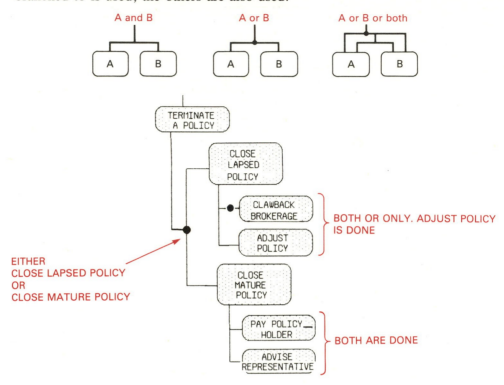

One-to-One and One-to-Many

Commonly, a facility is decomposed into other *single* facilities. An organization consists of *single* facilities. An organization consists of single suborganizations; a function consists of single subfunctions. Sometimes, however, we need to state that a facility is decomposed into subfacilities which occur multiple times. There is a one-to-many relationship between an activity and a component of that activity. This is shown using a crow's-foot to indicate one-to-many:

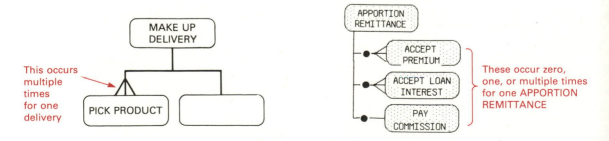

Sequence

Sometimes it is desirable to indicate in a decomposition diagram that processes occur in a given sequence. This is done with an arrow on the line to which the processes are connected. An arrow is used to indicate sequence. The sequence is normally from left to right (on a tree which progresses *down* the page) or from top to bottom (on a tree which progresses *across* the page).

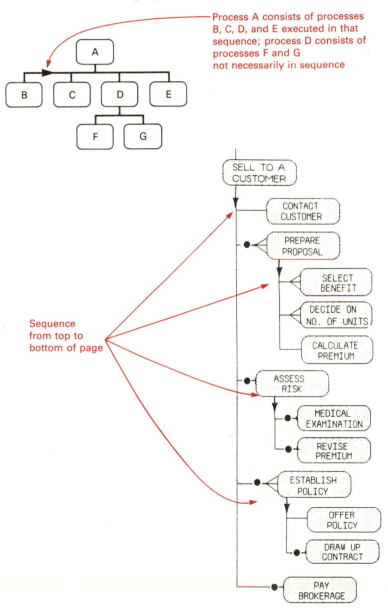

Process A consists of processes B, C, D, and E executed in that sequence; process D consists of processes F and G
not necessarily in sequence

Sequence from top to bottom of page

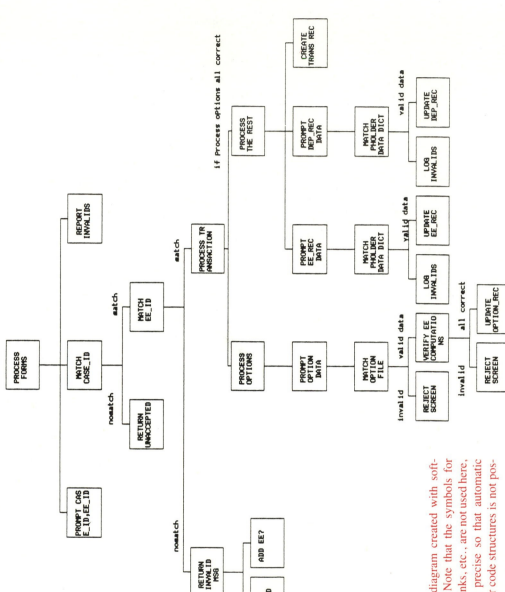

Figure 5.6 A decomposition diagram created with software on a personal computer. Note that the symbols for conditions, mutually exclusive links, etc., are not used here, which makes the diagram less precise so that automatic conversion to action diagrams or code structures is not possible.

ULTIMATE DECOMPOSITION

If an appropriate set of constructs are used, procedures can be decomposed until executable code is derived. This is particularly so with the code of some fourth-generation languages.

Decomposing all the way to executable code is called *ultimate decomposition*. To make it as easy as possible, we will use the technique of action diagrams described in Chapter 16.

REFERENCE

1. James Martin, *Strategic Data-Planning Methodologies*, Prentice-Hall, Inc., Englewood Cliffs, NJ, 1982.

6 DEPENDENCY DIAGRAMS

INTRODUCTION   A decomposition diagram shows how activities fit into a hierarchy. It does not show that certain activities are dependent on other activities. A time dependence exists between two activities if one cannot be carried out until the other has been completed. This is shown on a *dependency diagram*.

If process B cannot be performed until process A has been completed, we draw this as follows:

Dependencies among activities can apply to functions, processes, or procedures. We therefore have the following terms: function dependency diagram, process dependency diagram, and procedure dependency diagram.

THREE TYPES OF DEPENDENCIES   One activity may be dependent on another activity, for three types of reasons.

1. Resource Dependency

Activity A produces or modifies some tangible resource; activity B uses this resource. For example, DELIVER ORDER cannot occur before PICK GOODS, because there would be nothing to deliver. This type of dependency occurs only between resource-handling activities.

2. Data Dependency

Activity A creates or updates some data; activity B uses the data. For example, CREATE BACKORDER cannot occur until ACCEPT ORDER has occurred because CREATE BACKORDER needs certain data from the ACCESS ORDER process.

3. Constraint Dependency

An execution of some step in activity B depends on a constraint that was set in activity A, or the testing of a condition that was set in activity A. This type of dependency is to be avoided because it suggests tight coupling between activities. It should be replaced with a data dependency, so that the constraint or condition set in activity A results in data being passed to activity B. The activity boxes should be regarded as separate and disjoint with no entanglements between them except for the common use of data or tangible items.

FLOWS OR SHARED ACCESS

There are two types of interaction among dependent activities: flows and shared access to common storage. With resource dependencies, tangible goods may *flow* from one activity to another. Alternatively, the first activity may *put goods in a warehouse, and the dependent activity takes them from the warehouse.*

With data dependencies, data may *flow* from one activity to another. Alternatively, the first activity may *put data into on-line storage, and the dependent activity takes them from the storage*. Many dependent activities may share the same on-line files or data base.

INDEPENDENCE OF MECHANISMS

We have distinguished between *processes* and *procedures*. Diagrams with processes show *what* must happen to make the enterprise function, but not *how* this happens in terms of detail mechanisms. Diagrams with procedures show *how*, and are concerned with the mechanisms. We could call the former *mechanism-independent diagrams*. They are sometimes called *logical diagrams*, but this term has other connotations.

Mechanism-independent diagrams are not concerned with whether data flow directly from one process to another or whether they are passed via shared on-line storage. They enable us to diagram business processes without considering whether a data base is used, or even whether a computer is used. End users tend to relate well to mechanism-independent diagrams which illustrate their business processes.

As we descend into more detail we need to consider the use of data bases. At this stage in the design, dependency diagrams (or data flow diagrams, which are a specific form of dependency diagram), become inadequate. We need the greater detail of data navigation diagrams and action diagrams, discussed later.

CONSTRUCTS ON DEPENDENCY DIAGRAMS

As with other types of diagrams we can use the constructs of Chapter 4 to provide more information on dependency diagrams:

- Optionality
- Cardinality
- Branching
- Mutual exclusivity
- Loops
- Parallelism

Optionality

One process may give rise to another process only when certain conditions apply. We show optionality with a dot in a gap on the link between processes or procedures. The dot (an ''o'' for ''optional'') may be labeled with a condition:

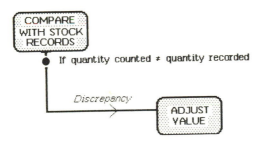

Often the condition is not written on the diagram because it is a high-level view of dependencies and the diagram needs to be uncluttered by detail.

A dot may be at either end of the line connecting blocks. Suppose that the line goes from process A to B. If the dot is by process A, the dependent process B may not occur. If the dot is by process B, process B may occur without process A having happened. Thus:

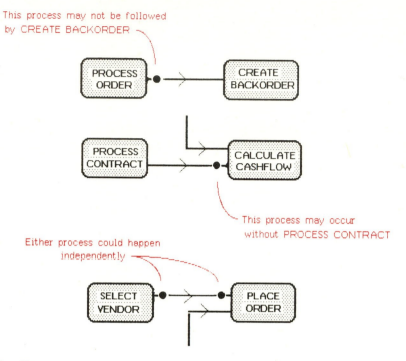

Cardinality

In the diagrams above the dependent process is executed *once* after the preceding process. In some diagrams we want to show that it may be executed multiple times:

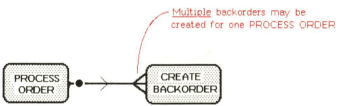

Similarly, a dependent process may follow multiple executions of a preceding process:

Less common is a one-to-many association at both ends of a link:

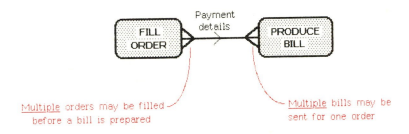

Branching

Sometimes one activity is dependent on many other activities. Lines from the preceding activity boxes join and enter the dependent activity box.

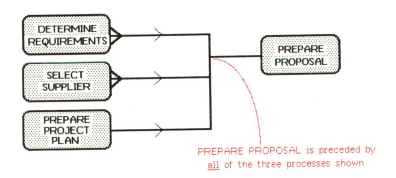

Conversely, one activity may give rise to many others:

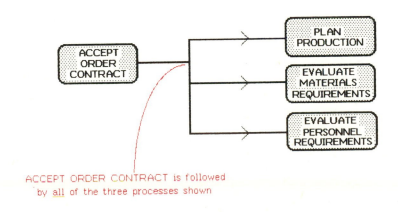

Mutual Exclusivity

Sometimes one or other of two activities must be performed, but not both. Sometimes one of several activities must be performed. These *mutually exclusive* choices of activity are shown by a branch with a dot in it—the "OR" dot used earlier:

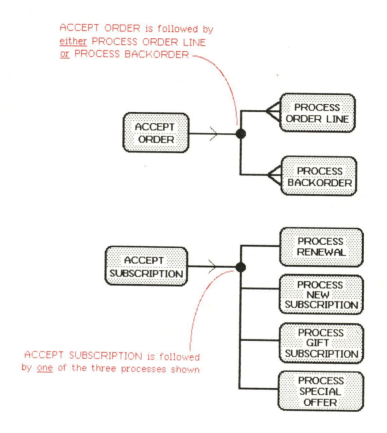

In the case above one activity is followed by alternative activities. We can also have one activity being dependent on alternative activities. For example, a depot may order stock of a particular product in three possible circumstances. One, a routine process MONITOR STOCK, indicates that the stock has dropped below a reorder level. Two, a CREATE BACKORDER process has been initiated because a customer order cannot be filled. Three, a delivery is being loaded and the PICK STOCK reveals unexpectedly that there is insufficient stock. No other process can cause more stock to be ordered. These three processes are independent:

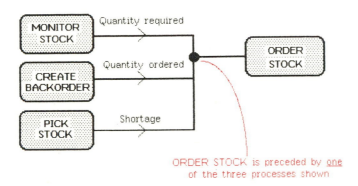

Loops

On rare occasions a process is drawn which depends on itself. Decomposing an order into a bill of materials may result in subassemblies having to be decomposed into lower-level subassemblies or components. Decomposing a complex project into activities may result in activities themselves being further decomposed:

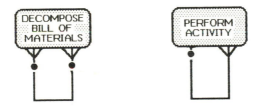

The diagram on the right above is, in effect, a tree structure where every block in the tree has the same label: PERFORM ACTIVITY. The diagram on the *left* above is, in effect, a network structure (plex structure) where every block is labeled DECOMPOSE BILL OF MATERIALS.

Parallelism

Occasionally, two or more links join the same two activity blocks. If these go in the same direction, this might indicate that the processes have been insufficiently or incorrectly decomposed. Links going in opposite directions between two processes occur in feedback loops or control mechanisms:

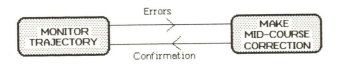

EVENTS

Some processes are triggered by other processes. This is often the case, but need not be so. Some are triggered by events. For example, the receipt of a payment may trigger a process. A process may be triggered by a customer telephoning to make a booking, a security alarm going off, the financial year ending, a bank's closing time being reached, a demand for information, and so on. They are all events external to the processes. We may talk about *event-triggered* processes and *process-triggered* processes.

A large arrow on a diagram is used to show that an event occurs:

This may be used on a dependency diagram, data flow diagram, state, or on other types of diagrams, such as a transition diagram.

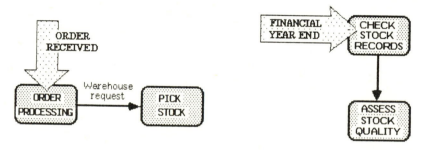

Some processes may be either event-triggered or process-triggered. For example, ALLOCATE PAYMENT in the following diagram may be triggered either by the event PAYMENT ARRIVES or by the process PRODUCE IN-VOICE:

The optionality dots before ALLOCATE PAYMENT indicate that this is *sometimes* triggered by PRODUCE INVOICE and *sometimes* by the event PAYMENT ARRIVES

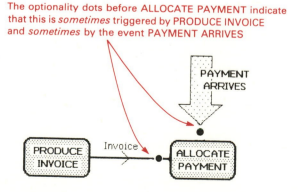

In some cases, both a preceding process and an event must occur before a given process takes place. In this case the event arrow links to the dependency arrow.

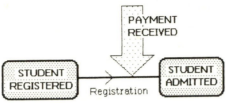

SEQUENCE   The sequence in which activities are carried out is indicated by the arrows on dependency diagrams. However, when multiple lines leave one block, the sequence in which they are executed is not clear unless they are numbered on the diagram. Similarly, when a branch occurs, the lines leaving the branch need to be numbered:

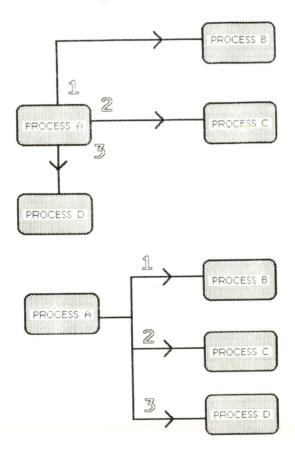

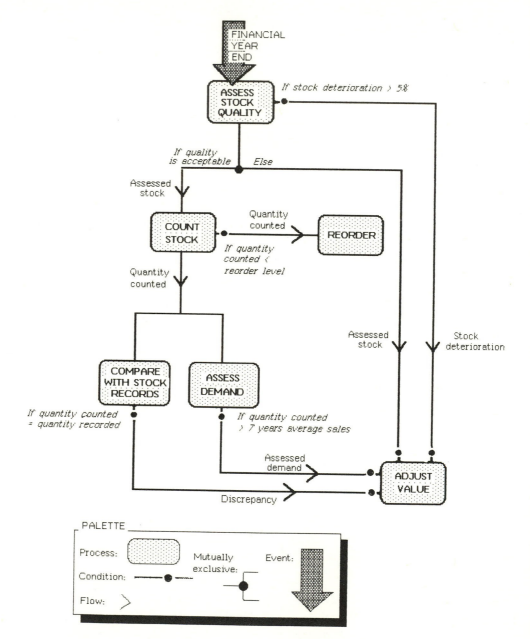

Figure 6.1 Dependency diagram.

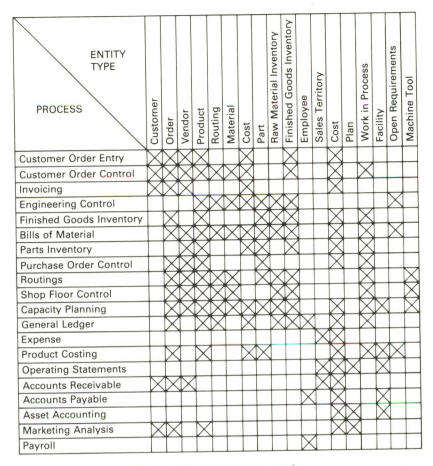

Figure 6.2   Data/process matrix.

When the lines are numbered in this way, we have sufficient information to convert the dependency diagrams into code structures with action diagrams.

CONVERSION TO ACTION DIAGRAMS   A dependency diagram can be converted automatically into an action diagram. An action diagram that shows sufficient detail can be thought of as being like a skeleton of a program. The skeleton is filled in to produce an executable program. This process can be done in a fast, computer-aided fashion. Chapter 16 discusses action diagrams in detail, and Fig. 16.11 shows illustrations of the automatic conversion of dependency diagrams to action diagrams.

USES OF DEPENDENCY DIAGRAMS

Figure 6.1 shows an example of a dependency diagram. By far the most commonly used form of dependency diagram is the *data flow diagram* discussed in the following chapter.

It is not practical or desirable to show every dependency among processes. A process may use data or resources that were created by another process months before. In order to sell goods it is necessary to hire salespersons, but these two activities would not normally be linked on a dependency diagram. Dependency diagrams are used to show processes with a close relationship that must be analyzed in order to understand how the enterprise functions.

There are often dependencies which are not drawn, between processes that use the same data. This shared usage of data-base records is illustrated on a matrix chart like that shown in Fig. 6.2. Figure 6.2 shows entity-record types and processes which create, retrieve, update, or delete the records. Although one process may use data that another process creates or updates, these processes may be sufficiently unconnected in time or sequence that they would not be combined on one dependency diagram.

7 DATA FLOW DIAGRAMS

INTRODUCTION   A data flow diagram is a commonly used form of dependency diagram. Unlike the dependency diagrams of Chapter 6, it shows data stores and external sources and destinations of data.

This chapter describes data flow diagrams as they are commonly drawn. Although they should, data flow diagrams usually do not show cardinality, optionality, mutual exclusivity, loops, or events as described in Chapter 6.

A data flow diagram shows processes and the flow of data among these processes. At a high level it is used to show business processes and the transactions resulting from these processes, whether paperwork or computer transactions. At a lower level it is used to show programs or program modules and the flow of data among these routines.

A data flow diagram is used as the first step in one form of structured design. It shows the *overall data flow through a system or program*. It is primarily a systems analysis tool used to draw the basic procedural components and the data that pass among them.

DEFINING DATA FLOW   Figure 7.1 is an example of a data flow diagram for an order acceptance procedure. In this procedure customer orders are processed. When an order is placed, the customer record is inspected to determine if the customer exists and whether the customer credit is in good standing. Then for each product on the order of a valid customer, product availability is checked to determine if the product is available and when it can be supplied. If there is insufficient quantity on hand, a backorder is created. Finally, an order confirmation is created and the customer order is completed with order status, order total amount due, and delivery information. The data flow diagram shows how data flow through a logical system, but it does *not* give control or sequence information.

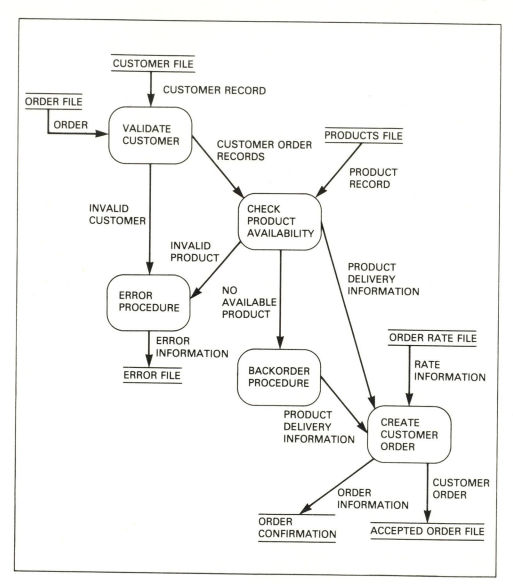

Figure 7.1   A data flow diagram for the order acceptance procedure shows the data flow and processes performed to handle customer orders.

COMPONENTS OF A DATA FLOW DIAGRAM

A *data flow diagram* (DFD) is a network representation of a system showing the processes and data interfaces between them [1]. The DFD is built from four basic components: the data flow, the process, the data store, and the terminator.

DATA FLOW

The *data flow* traces the flow of data through a system of processes. Direction of data flow is indicated by an arrow. The data are identified by name, with the name written alongside its corresponding arrow. For example:

PRODUCTS ORDERED ➤

In effect, the data flow shows how the processes are connected.

PROCESS

The *process* is a procedural component in the system. It operates on (or transforms) data. For example, it may perform arithmetic or logic operations on data to produce some result(s). Each process is represented by a circle or round-cornered box on the DFD. The name of the process is written inside the box. A meaningful name should be used to define the operation performed by the process. For example:

VALIDATE
CUSTOMER

No other information about what the process does is shown in the DFD.

Normally, data flow in and out of each process. Often there are multiple data flows in and out of a process. For example:

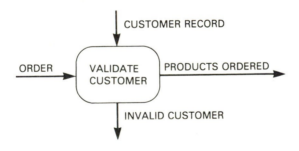

DATA STORE

A *data store* represents a logical file. It is drawn on the DFD as a pair of parallel lines, sometimes closed at one end. The name of the data store is written between the lines. For example:

PRODUCTS FILE   PRODUCTS FILE

Each data store is connected to a process box by means of a data flow. The direction of the data flow arrow shows whether data are being read from the data store into the process or produced by the process and then output to the data store.

In the example below, error information is produced by the ERROR PROCEDURE and written out to ERROR FILE:

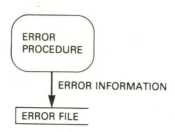

TERMINATOR

A *terminator* shows the origin of data used by the system and the ultimate recipient of data produced by the system. The origin of data is called a *source* and the recipient of data is called a *sink*. A rectangular box or double square, as shown below, is used to represent a terminator in a DFD:

Terminators actually lie outside the DFD.

LEVELING A DATA FLOW DIAGRAM

A DFD is a tool for top-down analysis. We can use DFDs to provide both high-level *and* more detailed views of a system or program [2]. What takes place within one box on a DFD can be shown in detail on another DFD.

Figure 7.2 shows the top-level view of the order entry procedure. It shows us that the order entry procedure is one of four procedures in the sales distribution system. But this view provides us with no detailed information about the processes and data flow needed to enter an order. If we need more detail, we must look inside the one process box to see what subprocesses are contained in the ORDER ENTRY PROCEDURE. This was the view provided by Fig. 7.1. Figure 7.3 gives an even more expanded view of this procedure by showing the subprocess contained in the VALIDATE CUSTOMER box.

We can continue to expand our view of the system by looking inside each process as far as we like. How detailed should our view be? A general rule of

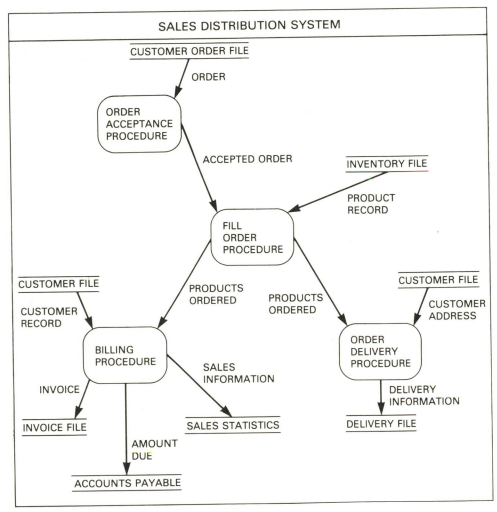

SALES DISTRIBUTION SYSTEM

CUSTOMER ORDER FILE

ORDER

ORDER
ACCEPTANCE
PROCEDURE

ACCEPTED ORDER

INVENTORY FILE

PRODUCT
RECORD

FILL
ORDER
PROCEDURE

CUSTOMER FILE

CUSTOMER
RECORD

PRODUCTS
ORDERED

PRODUCTS
ORDERED

CUSTOMER FILE

CUSTOMER
ADDRESS

BILLING
PROCEDURE

SALES
INFORMATION

ORDER
DELIVERY
PROCEDURE

INVOICE

DELIVERY
INFORMATION

INVOICE FILE

SALES STATISTICS

DELIVERY FILE

AMOUNT
DUE

ACCOUNTS PAYABLE

Figure 7.2   A top-level view of the order acceptance procedure shows that it is one of four subprocedures in the sales distribution system.

thumb is as follows: Keep expanding process boxes to create more detailed DFDs until operations performed by each process box can be described in a one-page specification [3].

At each level, the DFD should probably contain fewer than 12 process boxes and probably only six or seven. Larger DFDs are a sign of trying to show too much detail and are difficult to read. This process of defining a system in a top-down manner is called *leveling* a DFD.

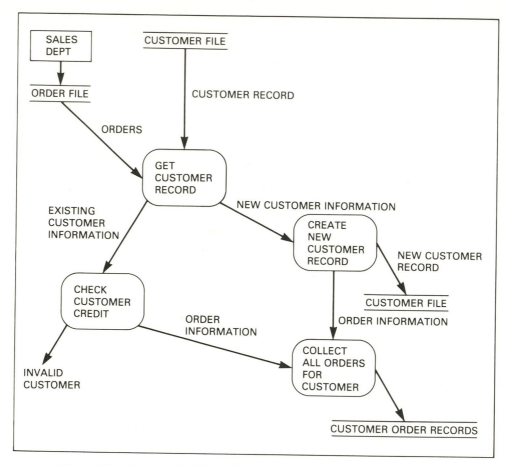

Figure 7.3 An expanded view of the VALIDATE CUSTOMER procedure shows four subprocedures: GET CUSTOMER RECORD, CREATE NEW CUSTOMER RECORD, CHECK CUSTOMER CREDIT, and COLLECT ALL ORDERS FOR CUSTOMER.

PROCESS SPECIFICATION AND DATA DICTIONARY

When a DFD is developed during structured analysis, a process specification and a data dictionary are also developed to give additional system information.

A *process specification* is created for each process box in the lowest-level DFD. It defines how data flow in and out of the process and what operations are performed on the data. A process specification is described with other techniques illustrated in subsequent chapters. Figure 7.4 shows a structured English version of what takes place in the labeled VALIDATE CUSTOMER in Fig. 7.1.

VALIDATE CUSTOMER

```
FOR EACH ORDER REQUEST:
    FIND MATCHING CUSTOMER RECORD;
    IF CUSTOMER FOUND
        IF CUSTOMER CREDIT IS IN GOOD STANDING
            COLLECT ALL PRODUCT ORDERS FOR THAT CUSTOMER;
            CREATE CUSTOMER_ORDER RECORD
        ELSE (CUSTOMER NOT IN GOOD STANDING)
            WRITE INVALID CUSTOMER ERROR MESSAGE
        ENDIF.
    ELSE (NEW CUSTOMER TO BE ADDED TO FILE)
        CREATE NEW CUSTOMER RECORD;
        COLLECT ALL PRODUCT ORDERS FOR THAT CUSTOMER;
        CREATE CUSTOMER_ORDER RECORD
    ENDIF.
```

Figure 7.4   A process specification shows what happens inside one DFD process box (at the lowest level of a leveled DFD). This specification, written in structured English, shows what happens in the VALIDATE CUSTOMER box of Fig. 7.1.

CUSTOMER FILE = [customer records]

The brackets [ ] indicate that in this example
CUSTOMER FILE is *iterations* of customer records.

customer-record = customer-name + customer-address
+ payment-information + outstanding-orders
+ cust-type

The + sign indicates that in this example a customer record is
a composite data item made up of a *sequence* of the data
items listed above.

cust-type = corp | individual

The vertical bar | indicates that cust-type is either a corporation
or an individual.

Figure 7.5   The data dictionary contains definitions of all the data in the DFD.

The *data dictionary* contains definitions of all data in the DFD. It can also include physical information about the data, such as data storage devices and data access methods. An example is shown in Fig. 7.5. The term "data dictionary" as used in conjunction with defining a DFD is *not* used in the same way as it is used in conjunction with data-base management systems.

GANE AND SARSON NOTATION

Gane and Sarson [4] adopted slightly different diagramming conventions for data flow diagrams from those popularized by Yourdon and De Marco. In some ways the Gane and Sarson notation is better. Figure 7.6 summarizes the two.

Gane and Sarson draw a process as a rounded rectangle; Yourdon and De

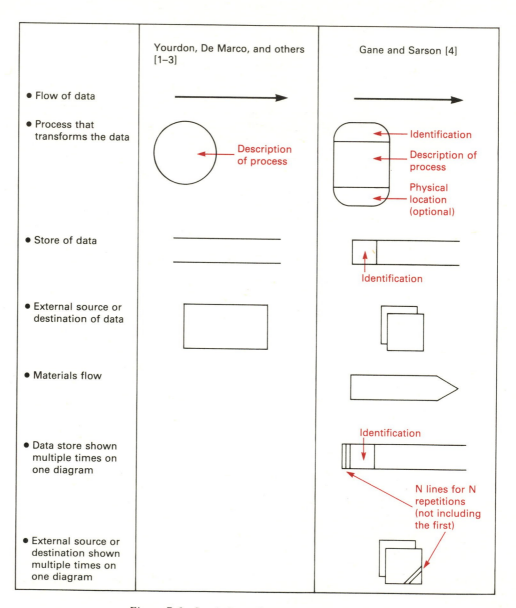

Figure 7.6 Symbols used on data flow diagrams.

Marco draw it as a circle. In computerized drawing it is easier to link *multiple* arrows to a round-cornered box than to a circle.

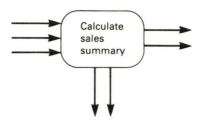

At the top of Gane and Sarson's block, a block number or other identifier is drawn:

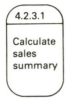

At the bottom the designer may optionally draw the physical location where the process takes place, or the name of the computer program that executes the process:

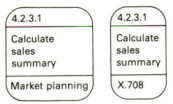

The data store is drawn with a block at the left that may give its number or identification. It may link it to a data model:

An external source or destination of data is drawn with a double square:

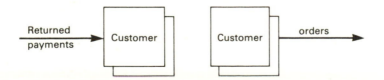

To simplify the mesh of lines on a drawing, a data store or external source or destination may appear multiple times. If a data store appears twice, a vertical line is drawn at the left side of its block. If an external block appears twice, a diagonal line is drawn in its bottom right-hand corner:

If either of these appears three times, two such lines are drawn:

If they appear N times, N − 1 such lines are drawn.

Data flow diagrams are useful for showing the flow of materials as well as computer data. It is important to indicate what computer data are and what they are not. Gane and Sarson use thick arrows to show the flow of materials:

Sometimes computer data accompany materials. Gane and Sarson draw the two together as shown in Fig. 7.7. Figs. 7.8 and 7.9 show typical Gane and Sarson data flow diagrams.

USE OF COMPUTER GRAPHICS

Data flow diagrams for complex projects become large, unwieldy, and difficult to maintain. The use of computer graphics solves these problems. When such tools are used, a developer creates diagrams on a workstation screen, updating the set of charts for the project which are available to all developers.

Various computerized versions of data flow diagramming exist. We will illustrate EXCELERATOR from InTech [5] and STRADIS/DRAW from MCAUTO

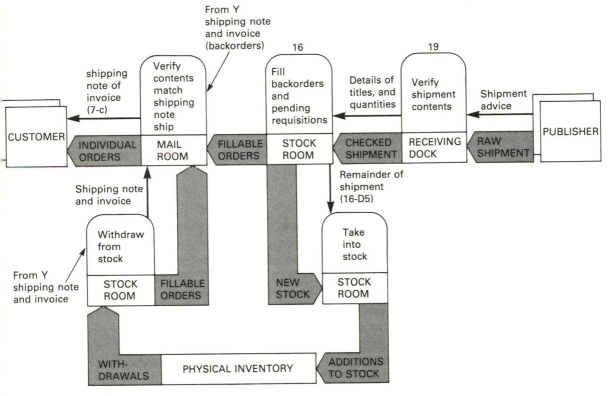

Figure 7.7   Gane and Sarson diagram showing physical flows (the heavy arrows) as well as data flows (the thin arrows).

[6]. The EXCELERATOR tool runs on personal computers. It supports various diagramming techniques and links them to a data dictionary. Fig. 7.8 shows a data flow diagram created with EXCELERATOR. The MCAUTO tool supports a version of the Gane and Sarson methodology called STRADIS; the graphics software for the methodology is STRADIS/DRAW. Figure 7.9 shows a data flow diagram created with STRADIS/DRAW.

The advantages of using such graphics tools are:

- Significant cost and time savings
- Reduction in work needed to redraft graphics during development
- Significant reduction of maintenance effort
- Elimination of proofreading and potential for error introduction on diagram updates
- Capability to produce very large data flow diagrams.

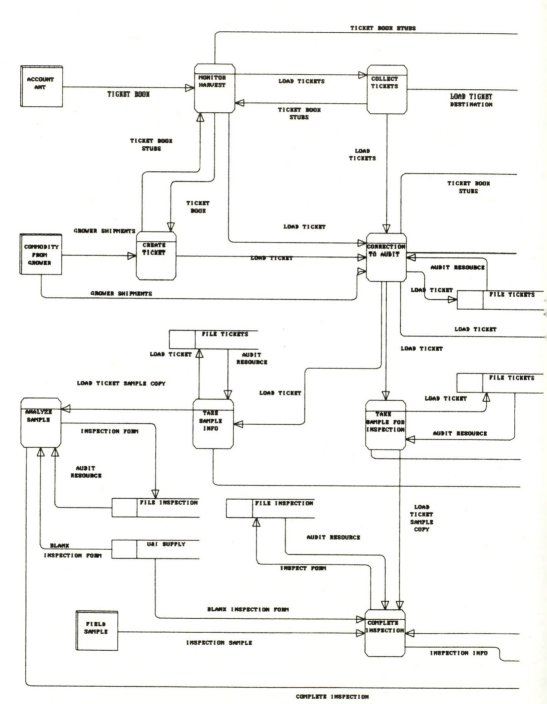

Figure 7.8 Data flow diagram produced and maintained on a personal computer with EXCELERATOR for In Tech [5] EXCELERATOR links data flow and other diagrams to its data dictionary.

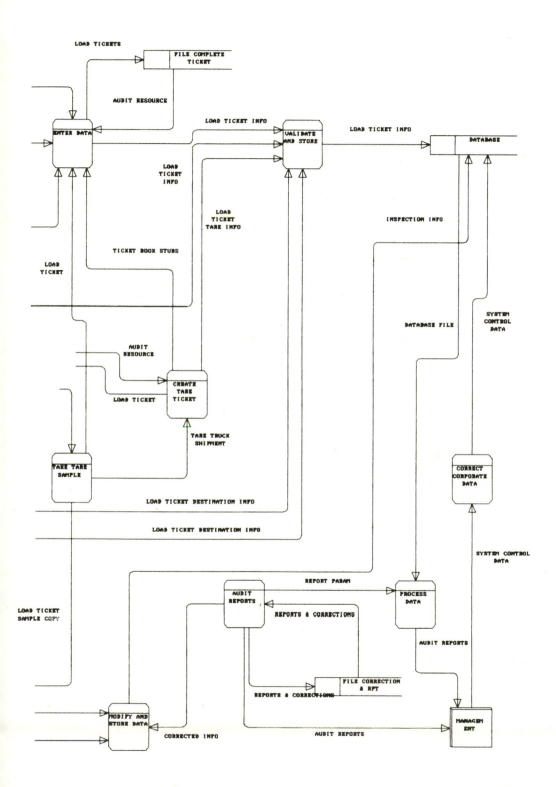

LOAD TICKETS

FILE COMPLETE
TICKET

AUDIT RESOURCE

ENTER DATA

LOAD TICKET INFO

VALIDATE
AND STORE

LOAD TICKET INFO

DATABASE

LOAD
TICKET
INFO

LOAD
TICKET
TARE INFO

INSPECTION INFO

TICKET BOOK STUBS

LOAD
TICKET

DATABASE FILE

SYSTEM
CONTROL
DATA

AUDIT
RESOURCE

CREATE
TARE
TICKET

LOAD TICKET

TARE TRUCK
SHIPMENT

TAKE TARE
SAMPLE

CORRECT
CORPORATE
DATA

LOAD TICKET DESTINATION INFO

LOAD TICKET DESTINATION INFO

SYSTEM CONTROL
DATA

LOAD TICKET
SAMPLE COPY

REPORT PARAM

AUDIT
REPORTS

PROCESS
DATA

REPORTS & CORRECTIONS

AUDIT REPORTS

FILE CORRECTION
& RPT

REPORTS & CORRECTIONS

MODIFY AND
STORE DATA

CORRECTED INFO

AUDIT REPORTS

MANAGEM
ENT

MCAUTO/IST

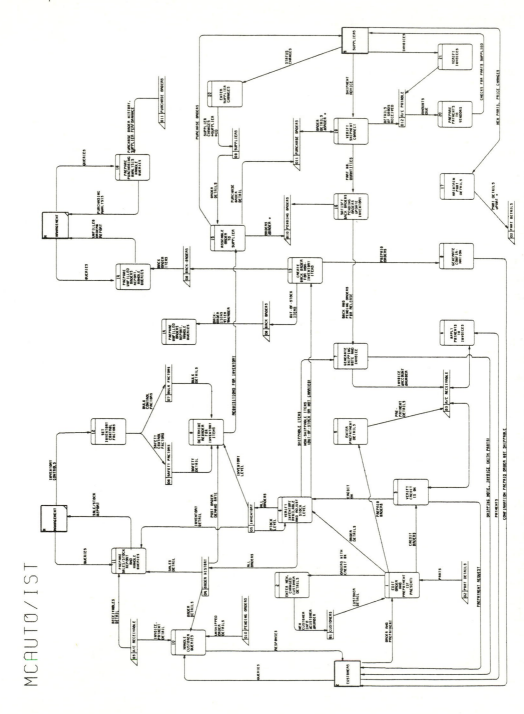

Figure 7.9 Data flow diagram created with STRADIS/DRAW. STRADIS is the MCAUTO version of the Gane and Sarson methodology [6].

106

COMMENTARY   Data flow diagrams are a very valuable tool for charting the flows of documents and computer data in complex systems. Some organizations extend their use beyond that into the internal structuring of programs. This use can be questionable, as we will discuss later; there are often better techniques.

In many cases we found data flow diagrams to be used badly. Large, complex specifications were created with the help of such diagrams, but the task of crosschecking all the inputs and outputs of data was not done adequately. The diagrams shown in this chapter are intended to illustrate this concern.

Figure 7.3 shows the VALIDATE CUSTOMER procedure. This procedure appears in Fig. 7.1. Figure 7.1, however, fails to show CUSTOMER FILE leaving the procedure. The VALIDATE CUSTOMER procedure of Fig. 7.3 creates a new customer record if the order is from a customer who does not yet exist. This is not reflected in Fig. 7.1.

The incompatibility between Figs. 7.1 and 7.2 is worse. The ORDER ACCEPTANCE PROCEDURE of Fig. 7.1 has entering it ORDER FILE, CUSTOMER FILE, PRODUCTS FILE, and ORDER RATE FILE, whereas the same procedure in Fig. 7.2 has only CUSTOMER ORDER FILE entering it. The outputs are similarly out of balance.

The authors did not make up these diagrams. We used diagrams drawn by a highly paid computer professional. When such discrepancies are pointed out, the charts seem surprisingly sloppy. But in practice we found this to be a common occurrence. Many sets of specifications, which programmers code from, have data flow diagrams on which the inputs and outputs are inconsistent. After the code has been written, it is expensive to sort out the resulting mess.

The authors were appalled to find the same types of errors in textbooks and courses. Readers might like to amuse themselves by checking the data flow diagrams in some of the popular books on structured methodologies.

In some cases the authors of data flow diagrams protest that when they drew the early, high-level diagrams, they could not yet know about the detail that would emerge when designing the later diagrams. When an analyst drew Fig. 7.2, for example, he thought merely of an ORDER ACCEPTANCE PROCEDURE working on a CUSTOMER ORDER FILE and producing a stream of ACCEPTED ORDERS. He had not yet invented the detail shown in Fig. 7.1. It is always true that the early diagrams are sketches, not yet detailed or precise. However, later, when the detail is worked out, it ought to be reflected back to the higher levels so that the higher levels become correct. In other words, the detail design of Fig. 7.3 should cause us to change Fig. 7.1, and the detail design of Fig. 7.1 should cause us to change Fig. 7.2.

With a computerized tool such changes can be made automatically. The detailed inputs and outputs of the lower-level layers can be reflected back to the higher-level layers. Amazingly, some computerized tools do not do this. They are merely drawing aids, not linked to the integrity checks that are desirable.

DATA LAYERING   The lower-level charts of a complex system often show, in total, many data items. If all of these data items were individually shown on the top-level charts, the charts would become very cluttered. The top-level charts may therefore show an aggregate data name which encompasses many of the lower-level names. The top-level chart may, for example, show MANUFACTURING DATA BASE as input; the next level shows JOB RECORD, which is part of the MANUFACTURING DATA BASE; the lowest level shows MACHINE—TOOL , START—TIME, and so on, which are part of the JOB RECORD. This is called *data layering*. We need drawing conventions to show how the data are layered. Different ways to draw structured data are shown in later chapters.

If detailed data flow diagrams are drawn without thorough understanding and structuring of the data, problems will result. Data analysis and data modeling, discussed later, need to go hand in hand with data flow diagramming.

REFERENCES

1. T. De Marco, *Structured Analysis and System Specification*, Yourdon Press, New York, 1978, p. 49.

2. E. Yourdon and L. Constantine, *Structured Design*, Yourdon Press, New York, 1978, pp. 38-40.

3. M. Page-Jones, *The Practical Guide to Structured Systems Design*, Yourdon Press, New York, 1980.

4. C. Gane and T. Sarson, *Structured Systems Analysis: Tools and Techniques*, IST, Inc., New York, 1977.

5. EXCELERATOR Product Description, InTech, 5 Cambridge Center, Cambridge, MA 02142.

6. STRADIS/DRAW Product Description, MCAUTO, McDonnell Douglas Automation Company, Box 516, Dept. K277, Saint Louis, MO 63166.

8 THREE SPECIES OF FUNCTIONAL DECOMPOSITION

INTRODUCTION   We have indicated that most structured design employs *decomposition*. A high-level activity is decomposed into lower-level activities; these are decomposed further; and so on. A tree structure shows the decomposition. We will use the term *functional decomposition* to refer to the decomposition of functions, processes, and procedures.

THREE LEVELS OF THOROUGHNESS IN FUNCTIONAL DECOMPOSITION   There are three different categories of functional decomposition—three separate *species*, as a botanist would say about trees.

Species I

The most common type of functional decomposition is a tree structure which relates to functions and not to the data which those functions use.

Species II

The second species shows the data types which are input and output to each function. This can be much more thorough, because if it is handled by computer, the machine can check that the data consumed and produced by each functional node are consistent throughout the entire structure.

Species III

The third species is still more thorough. It allows only certain types of decomposition, which have to obey precise rules that are defined by mathematical axioms. The resulting structure can then be completely verified to ensure that it is internally consistent and correct.

We are going to advocate thoroughness, not just because we want to avoid errors in program specifications, but because the thorough techniques have proven in practice to save much time and money in the long run. The thorough techniques will lead to a higher level of automation.

We discuss the first type of decomposition in Chapter 5. Here we examine species II and III decomposition.

SPECIES II FUNCTIONAL DECOMPOSITION

In species II functional decomposition, the functions are related to the data they use. In computing, a function is an algorithm that takes certain inputs and produces certain outputs. We represent it mathematically as:

$$y = F(x)$$

where

F is the function
x is the input(s)
y is the output(s)

Functional decomposition done as shown in Figs. 5.1 to 5.5 does not take the inputs and outputs into consideration.

We could draw the function $y = F(x)$ as follows:

$$y \boxed{\quad F \quad} x$$

where x is the input to a block labeled F which produces an output y. This type of diagram is the basis of the HOS notation, which we will discuss later. It relates neatly to mathematical notation but seems unnatural to the uninitiated because the input is on the right rather than on the left. It tends to lead to charts which are wide horizontally and thus difficult to print and manipulate. It is, for most people, more natural to use a vertical drawing with the input at the top and the output at the bottom:

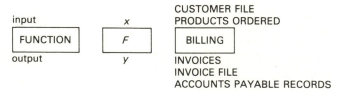

In Chapter 3 we advocated drawing a tree structure with square brackets when designing programs. Figure 8.1 shows the right-hand part of Fig. 5.4 drawn in this way. We can extend these brackets into a rectangle and show the data they use. This is done in Fig. 8.2; the input data of each function are written at its top right-hand corner, and the output at its bottom right-hand corner.

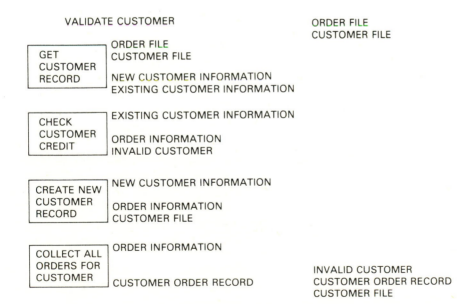

VALIDATE CUSTOMER

GET
CUSTOMER
RECORD

CHECK
CUSTOMER
CREDIT

CREATE NEW
CUSTOMER
RECORD

COLLECT ALL
ORDERS FOR
CUSTOMER

Figure 8.1　Hierarchical decomposition of the function VALIDATE CUS-TOMER. Figure 8.2 extends this diagram to show the data used.

VALIDATE CUSTOMER　　　　　　　　　ORDER FILE
　　　　　　　　　　　　　　　　　　CUSTOMER FILE

GET　　　　　ORDER FILE
CUSTOMER　　CUSTOMER FILE
RECORD
　　　　　　NEW CUSTOMER INFORMATION
　　　　　　EXISTING CUSTOMER INFORMATION

CHECK　　　　EXISTING CUSTOMER INFORMATION
CUSTOMER
CREDIT　　　ORDER INFORMATION
　　　　　　INVALID CUSTOMER

CREATE NEW　NEW CUSTOMER INFORMATION
CUSTOMER
RECORD　　　ORDER INFORMATION
　　　　　　CUSTOMER FILE

COLLECT ALL　ORDER INFORMATION
ORDERS FOR
CUSTOMER　　　　　　　　　　　　　INVALID CUSTOMER
　　　　　　CUSTOMER ORDER RECORD　CUSTOMER ORDER RECORD
　　　　　　　　　　　　　　　　　　CUSTOMER FILE

Figure 8.2　Hierarchy of Fig. 8.1 drawn to show the data inputs (top right-right corners) and outputs (bottom right-right corners).

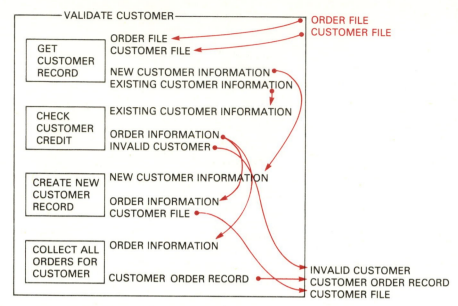

Figure 8.3 Figure 8.2, showing the passage of data among the functions. Figure 8.1 is the left-hand part of this diagram.

CHECKING THE USE OF DATA

We can now apply some checks. We can check that the inputs to VALIDATE CUSTOMER are all used by its internal blocks and that its outputs all come from its internal blocks. We can check that no internal block uses data that do not originate somewhere, and no internal block produces data that do not go anywhere. Figure 8.3 uses red arrows to show the passage of data among the functions.

Now let us work our way one step farther up the tree and show ORDER SERVICING, or at least that part of it represented by the three rightmost brackets of Fig. 5.4. Figure 8.4 shows this.

By checking all the transfers of data, we can be much more thorough than with simple functional decomposition. The chart, however, has become complex, and we have decomposed only one of the four blocks in ORDER SERVICING, and within that only one of the five blocks in ORDER ACCEPTANCE. If we decomposed all of the blocks to the same level, the chart would be about 20 times as large and complex. Getting it correct would strain the patience of a monk.

However, failing to get it correct is expensive. It means that our specifications are wrong. It is enormously cheaper to find errors at the specification stage than after the programmers have written code.

Full expansion and checking of the blocks in complex systems necessitates the use of a computer. The analysts should be able to build up the diagram a

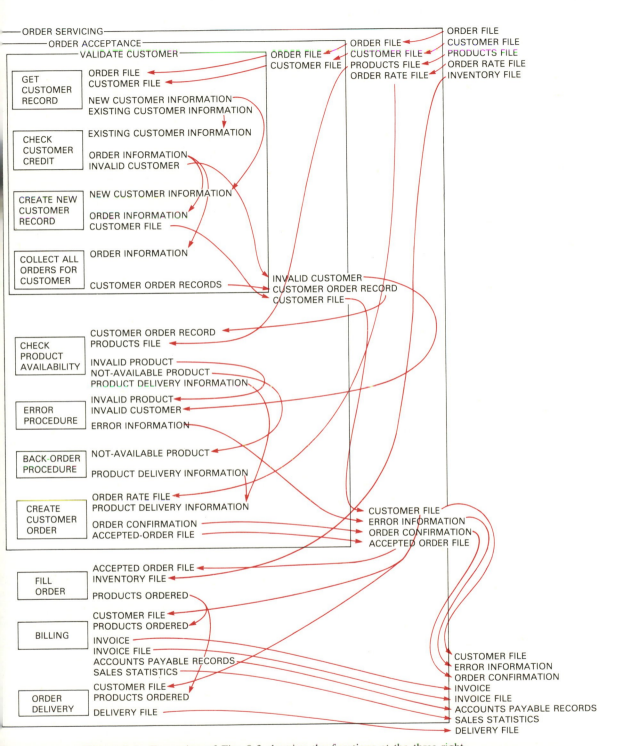

Figure 8.4 Expansion of Fig. 8.3 showing the functions at the three right-hand levels of Fig. 5.4.

113

step at a time, usually working from the highest level. They are likely to start with species I functional decomposition and then add the data. At the lower levels they will discover more details about the data required, and these details need to be reflected back up to the top. The computer can check the consistency and completeness of all the data transfers.

FUNCTIONAL DECOMPOSITION AND DATA FLOW DIAGRAMS

Figure 8.4 shows the same information as that shown in the data flow diagrams of Chapter 7. A functional decomposition diagram can be converted into a data flow diagram, and vice versa.

For some situations it is easier to think of system activities in terms of data flow. For others it is easier to think of it in terms of functions and their decomposition. Both of these when carried through to the level needed for program design become detailed, like Fig. 8.4, with 20 times as many blocks. Both therefore need computerized representation and checking. A computer graphics tool is needed which can relate the two and help analysts fill in the detail without inconsistencies.

Data flow diagrams tend to be more useful for showing the flow of documents in an organization or the way in which one business event triggers other events. They give a pictorial representation of the movement of tangible data, which the end users can relate to and be trained to draw and check. As systems analysis moves to the more detailed task of program design, hierarchical structures are more useful. Hierarchical structures will be converted into program code, as we will see later.

SPECIES III FUNCTIONAL DECOMPOSITION

With functional decomposition as normally practiced we can decompose a function any way that comes into our head. The third and most rigorous species of functional decomposition allows us to decompose in only certain ways which are defined with mathematically precise axioms.

In Fig. 8.3, for example, we have more than one type of decomposition. The first block of VALIDATE CUSTOMER is GET CUSTOMER RECORD. This is *always* performed. The second and third blocks are CHECK CUSTOMER CREDIT and CREATE NEW CUSTOMER. *Only one of these two is performed.* Which one depends on whether the order is from a new customer or from an existing customer.

The originators of the HOS methodology [1] concluded that all functional decomposition can be divided into *binary* decompositions. The parent function provides the input data for its binary children and receives the output. Three types of binary decomposition are needed:

- *Type i:*   The first function is executed; its results pass to the second function, which is then executed.

- *Type ii*:   *Either* the first function *or* the second function is executed.
- *Type iii*:   Both functions are executed independently.

Using these three binary primitives, the originators of the HOS methodology discovered that they could keep decomposing until program code could be generated automatically. Each binary decomposition follows rigorous mathematical axioms which enforce correctness so that the entire resulting structure can be proven to be internally correct. We thus have automatic generation of provably correct code [1].

The problem with this is that binary decompositions are so small that the overall design task is tedious. It is like trying to build a complex structure out of small Lego pieces. The solution is to design more powerful forms of decomposition which are themselves built from the primitives and are hence completely checkable. This is rather like building macroinstructions in programming.

Figures 8.5 to 8.7 show examples of HOS functional decomposition. In these diagrams, the functions are shown as rectangular boxes. The inputs to each box are on its right and the outputs are on its left. The type of decomposition is written on the line, which leaves the box, underneath it, going to the subordinate functions. The terms "COJOIN," "COOR," and "CONCUR" underneath the boxes are precisely defined types of decomposition that obey rules set by mathematical axioms. They are called *control structures*. The methodology is designed for computer graphics. As the designer builds diagrams such as Fig. 8.5 to 8.7 on the screen, the computer checks that the designer is obeying the axiomatic

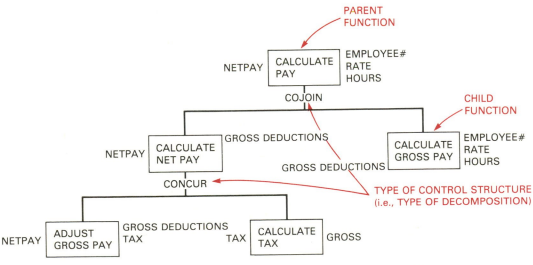

Figure 8.5 Functional decomposition done with HOS. Each decomposition must be of a defined type (e.g., COJOIN and CONCUR in this diagram) that obeys precise mathematical rules. In this diagram the input variables to each function are shown on its right and the output variables are shown on its left.

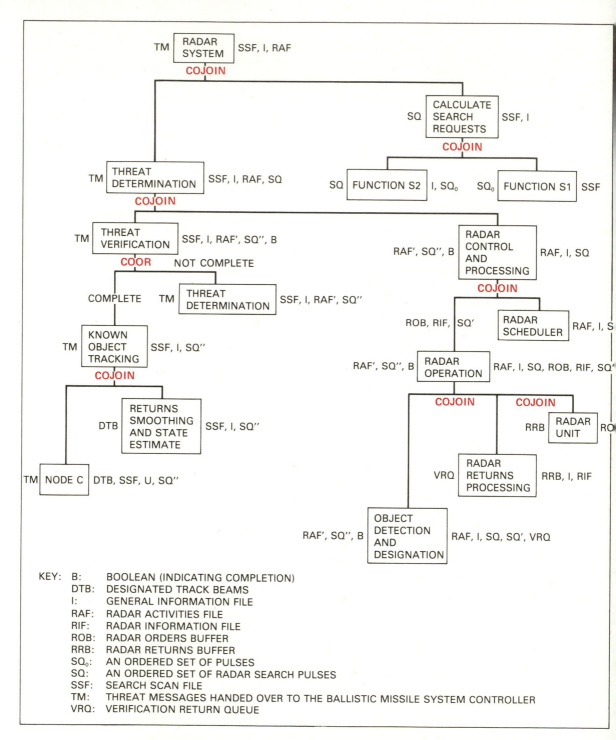

Figure 8.6 Overview of the radar system of a ballistic missile defense system, done with species III functional decomposition, which gives axiomatic checking of the decomposition [2,3].

116

rules. Precise details of the control structures (types of decomposition) are beyond the scope of this book. They are discussed in the senior author's book *System Design from Provably Correct Constructs* [1].

This more precise form of decomposition has several effects. First, it is built in a computer-assisted fashion step by step with the computer checking for syntax errors and periodically analyzing the chart to detect any errors in decomposition or use of data. With appropriately skilled analysts, complex specifications can be created much more quickly than with hand drawing.

Second, the specification that results is free of internal errors, ambiguities, omissions, and inconsistencies.

Third, the decomposition can be continued to a level of detail from which bug-free code is generated automatically.

Fourth, specifications can be built for extremely complex systems, which are free from internal errors, omissions, and inconsistencies.

Fifth, when changes have to be made, these are made on the terminal screen and all of the consequential changes that should result from any modification are indicated. The program can then be regenerated. Maintenance, in other words, is easier and faster.

Sixth, the technique seems alien and sometimes difficult to many traditional analysts and programmers. Bright analysts and new graduates often learn it very quickly, but much retraining is needed if a large DP team is to adopt it.

COMMENTARY   When the systems analysis profession has matured beyond its present stage we expect that most functional decomposition will progress to species II, with data usage being represented and analyzed using computerized tools. Much will progress to species III, with axiomatic control of the decomposition. Species III decomposition will probably progress far beyond that of HOS today, with a powerful library of tools and predesigned control structures.

REFERENCES

1. James Martin, "System Design from Provably Correct Constructs," Prentice-Hall Inc., Englewood Cliffs, NJ, 1985.

2. Wm. R. Hackler and A. Samarov, "An AXES Specification of a Radar Scheduler," Technical Report #23, Higher Order Software Inc., Cambridge, MA, Nov. 1979.

3. R. Hackler, "An AXES Specification of a Radar," Proceedings, Fourteenth Hawaii International Conference on System Sciences, Vol. 1, Honolulu, Hawaii, Western Periodicals Company, January 1981.

4. Details of USE-IT are available from HOS Inc., Cambridge, MA, Tel: (617) 661-8900.

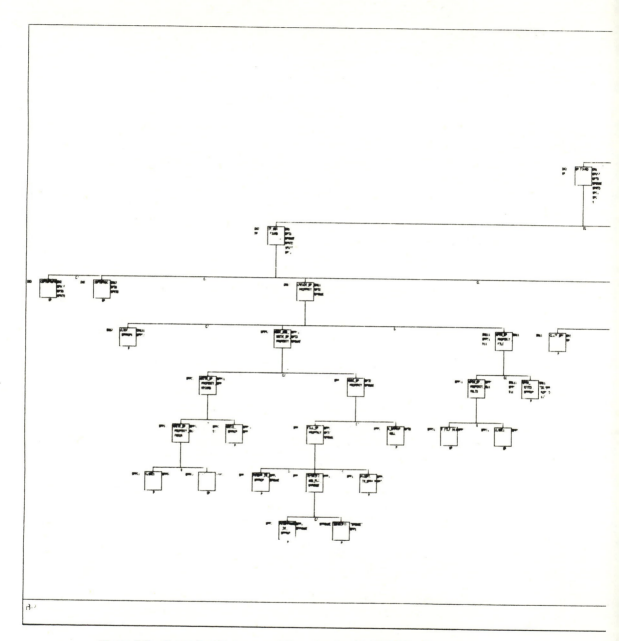

Figure 8.7 A species III decomposition plotted with USE-IT software from HOS Inc. [4]. Each data type usage and decomposition is verified mathematically by the software. The analyst normally windows around such a chart at his workstation, seeing about six blocks at a time.

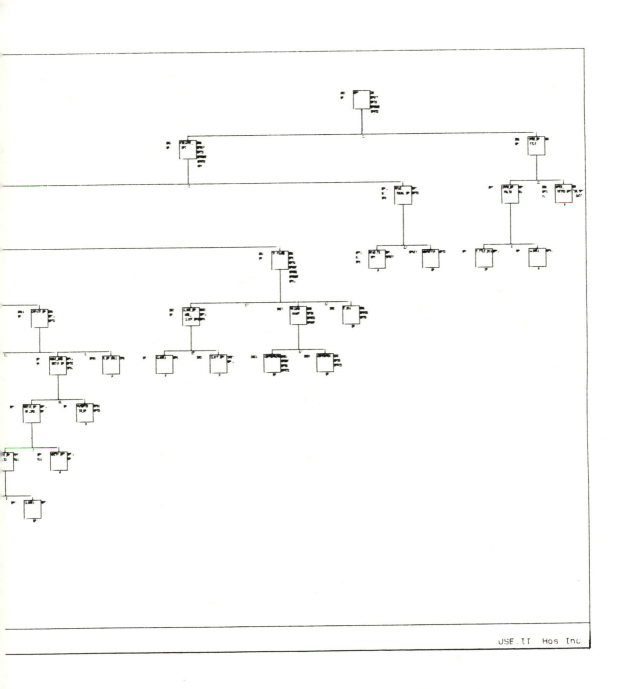

119

STRUCTURE CHARTS

9

HIERARCHICAL DIAGRAM

Structure charts are a form of functional decomposition. Together with data flow diagrams they constitute the structured design methodology in common use at the time of this writing [1]. The basic building block of a program is a module. Structured programs are organized as a hierarchy of modules. The *structure chart* is a tree or hierarchical diagram that defines the overall architecture of a program by showing the program modules and their interrelationships.

Figure 9.1 shows the structure chart for a subscription system. The purpose of the system is to process subscription transactions against a subscription master file. There are three types of transactions: new subscriptions, renewals, and cancellations. Each transaction is first validated and then processed against the master file. For new subscriptions, a customer record is built and a bill is generated for the balance due. For renewals, the expiration date is updated and a bill is generated for the balance due. For cancellations, the record is flagged for deletion and a refund is issued.

The program is represented as a set of hierarchically ordered modules. Modules performing high-level program tasks are placed at the upper levels of the hierarchy, while modules performing low-level, detailed tasks appear at lower levels. Looking down the hierarchy, the modules at each successive level contain tasks that further define tasks performed at the preceding level. For example, the task GET VALID SUB ITEM is composed of two subtasks: READ SUB ITEM and VALIDATE SUB ITEM.

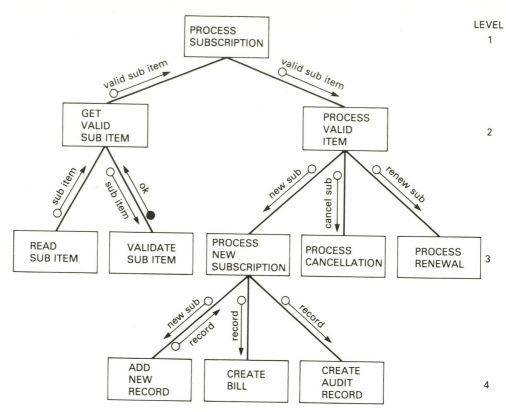

Figure 9.1 The structure chart defines the overall architecture of a program by showing the basic procedural components and their interrelationships. The arrows with a circle, ○────→, show data passing between the process blocks. The arrows with a dot, ●────→, show control information passing between the process blocks.

COMPONENTS OF A STRUCTURE CHART

The basic building blocks of structure charts are rectangular boxes and arrows connecting the boxes. Although the smallest possible structure chart contains only one box, most structure charts contain many boxes and connecting arrows.

Each rectangular box in the structure chart represents a module. Logically, a module is one problem-related task that the program performs, such as ADD NEW RECORD or CREATE BILL. Physically, a module is implemented as a sequence of programming instructions bounded by an entry point and an exit point. The module name is written inside the box:

```
┌──────────────┐
│ CANCEL       │
│ SUBSCRIPTION │
└──────────────┘
```

A descriptive name should be chosen to explain the task the module performs. Other than the module name, the structure chart gives *no* information about the internals of the module.

CONTROL RELATIONSHIPS

Modules are interrelated by a control structure. The structure chart shows the interrelationships by arranging modules in levels and connecting the levels by arrows.

An arrow drawn between two modules at successive levels means that at execution time program control is passed from one module to the second in the direction of the arrow. We say that the first module *invokes* or *calls* the second module. For example, in Fig. 9.1 module PROCESS NEW SUBSCRIPTION invokes the module CREATE BILL. After the module CREATE BILL finishes executing, control is returned to PROCESS NEW SUBSCRIPTION. However, we cannot tell by looking at the structure chart whether module PROCESS NEW SUBSCRIPTION invokes CREATE BILL one time or many times or what conditions, if any, are tested to make this invocation decision.

It is possible for one module to invoke multiple modules (or to have multiple sons). In the example below, module A has two sons: B and C. Since the structure chart does not show sequence, we do not know in what order a module invokes its sons. A module that has no sons is called a *leaf*.

The rules governing the program control structure are listed in Box 9.1.

BOX 9.1   Control rules for a structure chart

- There is one and only one module at the top (level 1) of the structure chart. This is where control originates. This module is called the *root*.

- From the root, control is passed down the structure chart level-by-level to the other modules. Control is always passed back to the invoking module. Therefore, when the program finishes executing, control returns to the root.

- There is at most one control relationship between any two modules in the structure chart. This means that if module A invokes module B, then module B *cannot* also invoke module A. Also, a module cannot invoke itself. (The reason for this restriction is that the principle of abstraction as implemented by levels of abstraction does not allow lower-level modules to be aware of the existence of upper-level modules.)

COMMON MODULES

It is possible for more than one module to transfer control to the same module. As shown below, module B and module C both invoke module E:

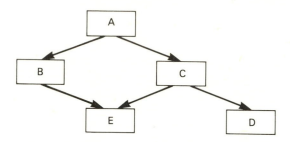

In this case, module E is called a *common module*. The diagram above is no longer a tree structure. It is drawn in this way to avoid having to show module E twice. A common module could itself have a family, as does CALCULATE DEDUCTIONS in Fig. 9.2.

LIBRARY MODULES

In some cases, the program developer may use pre-defined library modules. This is indicated in the structure chart by a rectangular box with double vertical lines:

Normally, a library module will appear as a leaf on the structure chart.

DATA TRANSFER

When control is transferred between two modules, data are usually transferred as well. Data may be transferred in either direction between the modules. Direction is shown by drawing a small arrow. Data names are written beside the arrow to identify what data are passed.

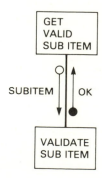

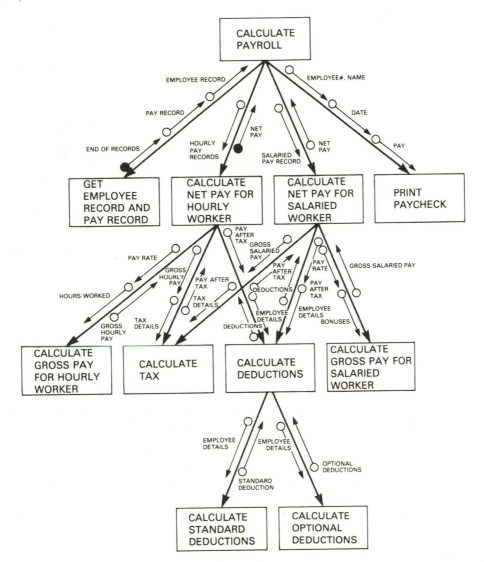

Figure 9.2 (Simplified) structure chart for a payroll program showing two common modules. The data labels on this style of diagram often become too cluttered.

There are two basic types of information that can be communicated between modules: data and control information [1]. Data comprise information used in the problem, such as Subscription Item. This type of information is identified by an arrow with an open circle as its tail, ○——→ . It is called a *data couple*. Control information is used by the program to direct execution flow, such as an error flag or end-of-file switch. It is identified by a dot at its tail, ●——→ . It is called a *control couple*, or sometimes a *flag*. Figures 9.1 and 9.2 show data couples and control couples passing.

Structure charts show small numbers of data types passing. They are not used to show large numbers, as with the outer block of Fig. 8.4. The little arrows attached to vertical links would be inappropriate for this. Even with small numbers the chart becomes cluttered, as in Fig. 9.2. If there are more than two or three data items passing from one block to another, this usually indicates that the structure chart should be further decomposed.

Although structure charts do show data passing between blocks, we would not class them a species II functional decomposition as described in Chapter 8 because the data inputs and outputs of an integrated hierarchy are not automatically analyzed.

SEQUENCE SELECTION AND ITERATION

Three important control structures in program design are sequence, selection, and iteration. *Sequence* refers to the order in which blocks are executed. *Selection* refers to the use of conditions to control whether or not a module is executed, or which of several blocks are executed. *Iteration* refers to the control of loops.

Structure charts do not show these. They are regarded as detailed internals of the blocks which are shown by a different technique, usually pseudocode (Chapter 14).

It might be thought that the sequence of executing the modules could be shown by the left-to-right sequence of the blocks. Common modules can make this difficult. Yourdon and others stress that the module drawing is independent of sequence of execution.

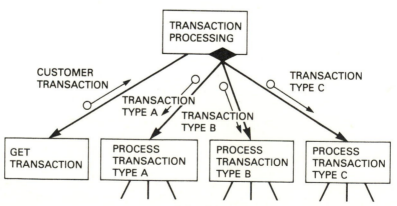

Figure 9.3 The black diamond in the top block is called a *transaction center*. Separate modules are used for processing each type of transaction. The transaction center determines the type of a transaction and transfers control to the appropriate module.

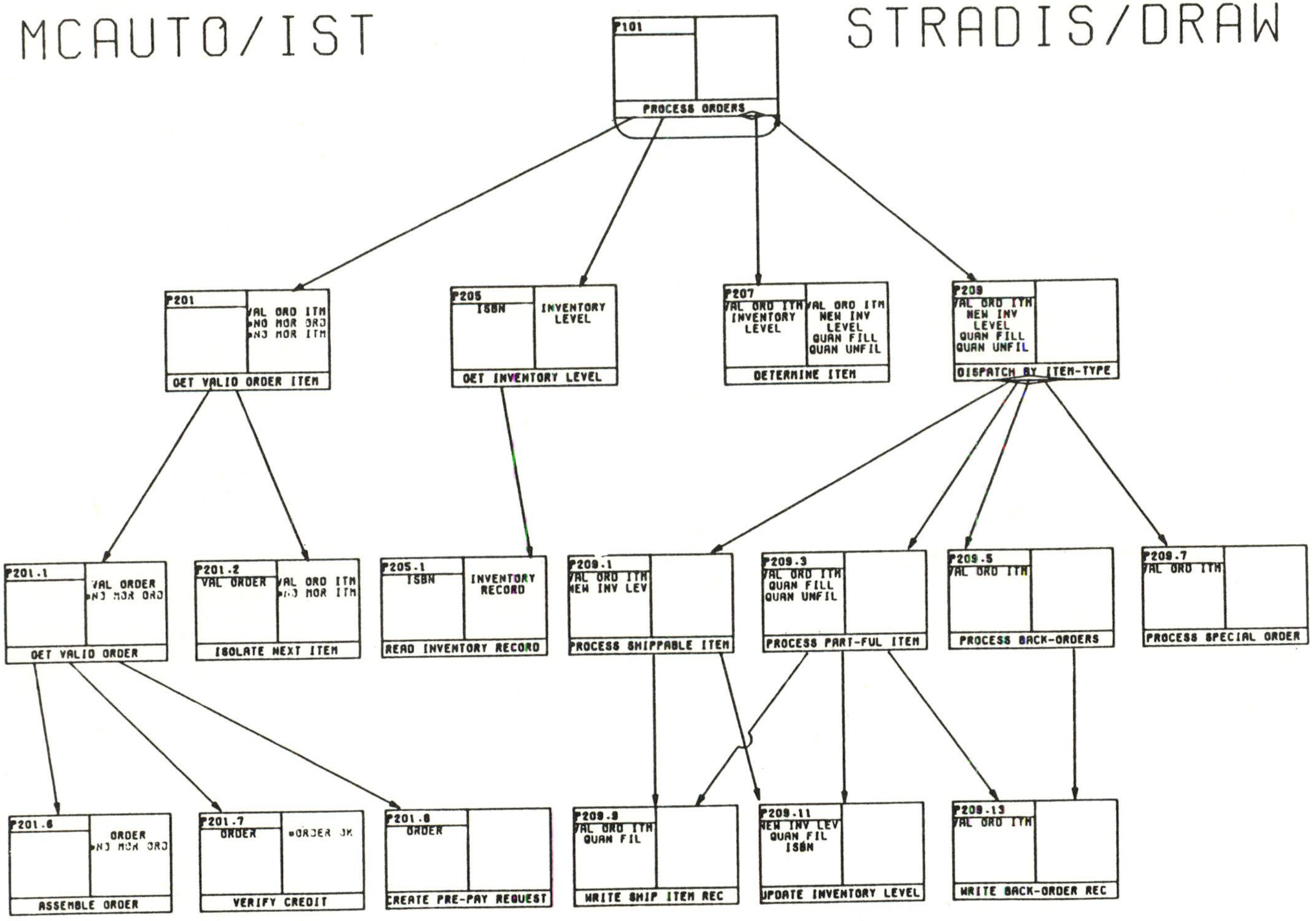

Figure 9.4 Data structure diagram drawn with MCAUTO's STRADIS/DRAW. Data transfers are shown as in Fig. 9.5.

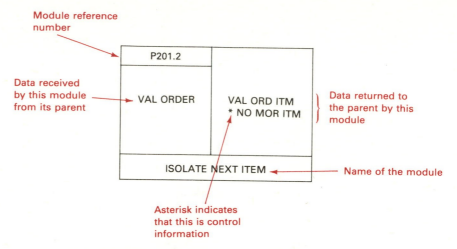

Figure 9.5   With MCAUTO's STRADIS/DRAW (Fig. 9.4), data transfers are placed inside the rectangle instead of being indicated with arrows as in Fig. 9.2.

TRANSACTION CENTER

Where multiple types of transactions are processed, a separate program module may be used for each transaction type. The allocation of control to these separate modules may be shown on a structure chart by a black diamond, as shown on Fig. 9.3. This black diamond could have been used in the block labeled DISPATCH in Fig. 9.1 for splitting the control among NEW SUBSCRIPTION, CANCEL SUBSCRIPTION, and RENEW SUBSCRIPTION.

COMPUTER GRAPHICS

The STRADIS/DRAW software mentioned in Chapter 7 provides an interactive graphics tool for creating, editing, and maintaining structure charts. Figure 9.4 shows a version of a structure chart created with this tool [2].

Because the arrows indicating data transfers become cluttered on structure charts such as Fig. 9.2, STRADIS/DRAW puts the data transfers inside the rectangles, as shown in Figure 9.5. Control flags are marked with an asterisk as is "NO MOR ITM" in Fig. 9.5. Even with this approach the data transfers can be cluttered and their names overly abbreviated.

EXCELERATOR, from InTech [3], enables structure charts and other types of diagrams to be drawn on the screen of a personal computer and linked to a dictionary. Data flow diagrams are drawn using the same dictionary. Fig. 7.8 showed a data flow diagram drawn with EXCELERATOR; Fig. 9.6 shows a structure chart.

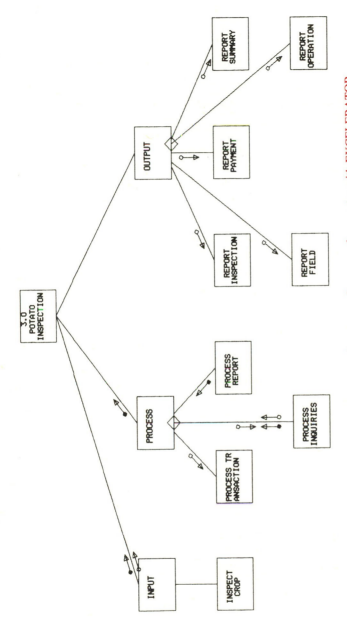

Figure 9.6 A structure chart drawn on the screen of a personal computer with EXCELERATOR, from InTech [3]. The items on the diagram are in the EXCELERATOR dictionary. The designer can record and display details of each item.

REFERENCES

1. E. Yourdon and L. Constantine, *Structured Design*, Yourdon Press, New York, 1978, pp. 42–50.

2. *STRADIS/DRAW Reference Manual*, MCAUTO, McDonnell Douglas Automation Co., Box 516, Saint Louis, MO 63166.

3. EXCELERATOR Reference Manual, InTech, 5 Cambridge Center, Cambridge, MA 02142.

10 HIPO DIAGRAMS

HIPO DIAGRAMS   HIPO stands for Hierarchical Input–Process–Output. A *HIPO diagram* is a diagramming technique that uses a set of diagrams to show the input, output, and functions of a system or program. It can give a general or detailed view of a system (or program).

Like a structure chart, HIPO shows *what* the system does, rather than *how*. There are three basic kinds of HIPO diagrams:

- Visual table of contents
- Overview diagrams
- Detail diagrams

Figures 10.1, 10.2, and 10.3 illustrate these.

The *visual table of contents* looks very similar to a structure chart. An example of a visual table of contents for the subscription system is shown in Fig. 10.1. The purpose of this system is to process three types of subscription transactions: new subscriptions, renewals, and cancellations.

Each box in the visual table of contents can represent a system, subsystem, program, or program module. Its purpose is to show the overall functional components and to refer to overview and detail HIPO diagrams. Notice, however, that it does *not* show the data flow between functional components or any control information. It does not show the arrows ○⟶ and ●⟶ which are on structure charts such as that shown in Fig. 10.1. Also, it does not give any information about the data components of the system (program).

Overview and *detail HIPO diagrams* give more information on each functional component shown in the visual table of contents. The distinction between them is the amount of detail they show. Overview diagrams describe the input, process, and output of the major functional components. According to IBM, their purpose is to provide "general knowledge of a particular function" [1]. On the

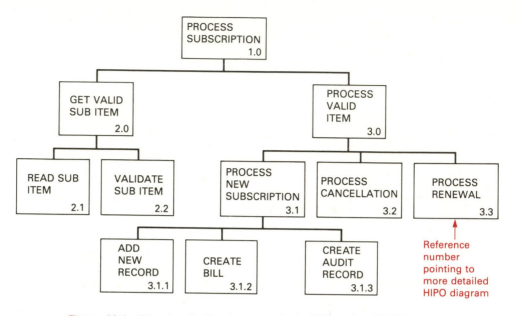

Figure 10.1   The visual table of contents is the highest-level HIPO diagram. It shows the overall functional components of a system or program. It does not give any control information, nor does it describe any data components, as does the equivalent structure chart of Fig. 9.1.

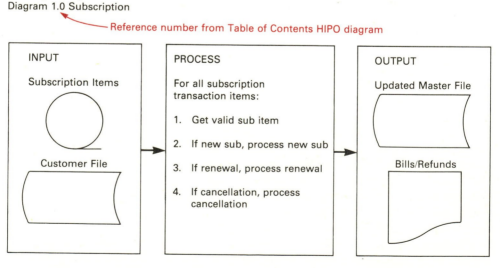

Figure 10.2   An overview HIPO diagram gives general information about the inputs, process steps, and outputs of one particular functional component in a system (program).

Subscription System
Diagram 2.2.2 Validate New Sub

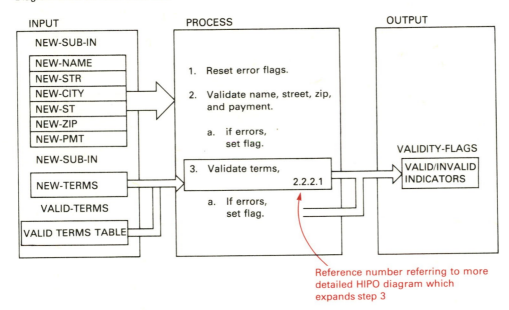

Figure 10.3 A detail HIPO diagram provides the information necessary to understand the inputs, processing steps, and outputs for a functional component. It represents the program design and can easily be transformed into program code.

other hand, detail diagrams provide ''all the information necessary to understand the function specified in the next higher [level] diagram'' [1]. Overview diagrams typically are used to describe top-level components in the visual table of contents, whereas detail diagrams are used to describe the low-level components. Figure 10.2 is the overview HIPO diagram for the PROCESS SUBSCRIPTION function in the subscription system. Figure 10.3 is the detail diagram for the VALIDATE NEW SUB function in the subscription system.

DIAGRAM COMPONENTS

Overview and detail HIPO diagrams look very much alike. They both consist of three parts: an input box, a process box, and an output box.

In a HIPO diagram, the left-hand box is the input portion of the diagram. It shows the input data items, which may be a file, a table, an array, or an individual program variable. Flowchart device symbols are used in overview diagrams to indicate the type of physical device, when it is known. For example, Fig. 10.2 shows that subscription transactions, which are one input to PROCESS SUB-

SCRIPTION, reside on a tape file and that the subscription master file, which is an output of PROCESS SUBSCRIPTION, is a disk file.

The center box is the process portion of the diagram. It contains a list of the process steps to be performed. These steps correspond closely to the subfunctions that were identified in the visual table of contents. For example, in Fig. 10.2, PROCESS SUBSCRIPTION has four process steps, each of which corresponds to the four subfunctions shown in Fig. 10.1. Also, control information is included to indicate the logic that governs the execution of the process steps. Figure 10.2 shows that all four process steps are performed for each transaction item on the subscription transaction tape.

The right-hand box is the output part of the diagram. It shows the output items produced by the process steps. Like an input data item, an output data item may be a file, a table, or a variable.

The three parts of the diagram are connected by arrows to show which input and/or output data items are associated with each process step. For example, in Fig. 10.3 New-Terms and Valid Terms Table are input to process step 3: Validate Terms. The Valid/Invalid Indicators are the output of process steps 2 and 3.

While the visual table of contents is similar to a structure chart, overview and detail HIPO diagrams are similar to a data flow diagram. Like a data flow diagram, they show the flow of data through processes. However, HIPO diagrams are more difficult to draw than are data flow diagrams. HIPO diagrams often require more verbiage and symbols to give the same information as that on a comparable data flow diagram.

ANALYSIS AND DESIGN TOOLS

HIPO diagrams are used as both analysis and design tools. At the analysis stage, they are used by the systems analyst to define in general the system (program) components. This general definition then becomes the starting point for system (program) design. During the design stage, the designer draws more detailed HIPO diagrams to describe each procedural component.

HIPO diagrams are hierarchical and can be used to describe the top-down programming process. During the top-down process, the process steps in higher-level HIPO diagrams are expanded into a set of lower-level HIPO diagrams. When the design process is finished, there will be one HIPO diagram for each functional component in the program. These HIPO diagrams are then translated into executable code during the implementation step. After implementation, they are retained as system documentation.

COMMENTARY

HIPO diagrams represent another type of procedure design tool. Although HIPO diagrams are a design aid, HIPO diagramming by itself is not a complete design methodology. It does not include any guidelines, strategies, or procedures to guide the analyst in building a functional specification or the designer in building a system or program design.

For smaller programs, HIPO diagrams may be a sufficient tool for program design. However, they quickly become cluttered and difficult to read when there are several process steps or input/output data items to show.

At the general level, a structure chart is usually preferred over a visual table of contents. Although both diagrams provide a hierarchical representation of the basic functional components, only the structure chart shows how the components are interrelated via data. At the detailed level, pseudocode is often preferred over HIPO diagrams because it provides more information in a more compact form, although a HIPO diagram shows input and output more clearly. Showing input and output and how they relate to procedural steps is a strength of HIPO diagrams which some other structured techniques do not have.

HIPO diagrams have no symbols for representing detailed program structures such as conditions, case structures, and loops. Narrative description is used for this. In addition, HIPO diagrams cannot represent data structure or linkage to data models.

REFERENCE

1. *IBM HIPO, A Design Aid and Documentation Technique*, GC20-185D, IBM Corp., White Plains, NY, 1974.

11 WARNIER–ORR DIAGRAMS

INTRODUCTION   Warnier–Orr diagrams are an alternative to HIPO diagrams or structure charts [1]. They are named after their two principal proponents, Jean-Dominique Warnier and Ken Orr. Like HIPO diagrams and structure charts, they aid the design of well-structured programs. They have certain advantages over these other structured methods. They are easy to learn and to use because they are composed of only four basic diagramming techniques.

A Warnier–Orr diagram represents graphically the hierarchical structure of a program, a system, or a data structure. It draws it horizontally across the page with brackets, instead of down the page with blocks, as shown in Fig. 11.1.

Each bracket in the diagram represents a functional breakdown of the item at the head of the bracket. This is similar to a structure chart or a visual table of contents HIPO diagram. In addition to showing hierarchical structure, a Warnier–Orr diagram also shows flow of control through the structure. This is similar to an overview or detail HIPO diagram.

REPRESENTATION OF DATA   Warnier–Orr charts can represent hierarchical data structures or reports as well as program structures. Figure 11.2 is an example of a Warnier–Orr diagram of an employee file. The diagram is read from left to right and from top to bottom within a bracket. The brackets enclose logically related items and separate each hierarchical level. The items are listed vertically. A meaningful name is given to each item. The number of times that an item occurs is written in parentheses below its name. For example, in Fig. 11.2, employee file occurs once. Within the file body, EMPLOYEE RECORD occurs from 1 to E times, as indicated by (1,E).

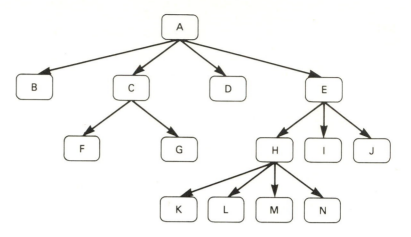

Figure 11.1 A decomposition diagram, structure chart, or HIPO diagram draws a hierarchical structure down the page with blocks; a Warnier–Orr chart draws it across the page with brackets.

If an item had the notation (0,1) under it, that would mean that it was either present or not. Warnier refers to this as an *alternation structure*. The alternation structure can be used with an "or" or "exclusive or" structure. If two items are separated by a "+", it means that one or the other or both items are included (i.e., or structure). If two items are separated by a "$\oplus$", it means that one or the other but not both items are included (i.e., exclusive or). For example, in the employee file shown in Fig. 11.2, an employee is either salaried or hourly, but not both.

Figure 11.3 is a sales contract and Fig. 11.4 is the Warnier–Orr diagram showing the data items included in this contract. Notice that not all the contract information is required, and that the order information may be repeated. There may be 1 to N sets of order information in one sales contract.

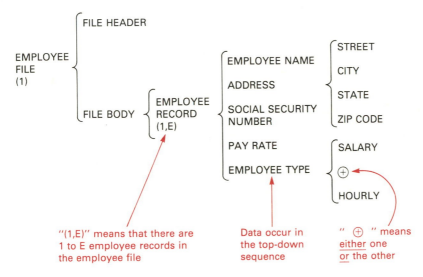

Figure 11.2 Warnier–Orr representation of an employee file as a hierarchical structure. The parentheses under an item indicate the number of times it is repeated. For example, there is 1 employee file; there are 1 to E employee records within the body. The ''⊕'' between SALARY and HOURLY means *either* SALARY *or* HOURLY, but not both (mutually exclusive selection). '' + '' would mean either one or the other or both (nonexclusive selection).

THE HOUSE OF MUSIC INC.

SALES CONTRACT

A Collins Corporation

Main Office
108 Old Street, White Cliffs, IL 67309
063 259 0003

Contract No. 7094

SOLD BY	Mike	DATE	6/10/83

Name: Herbert H. Matlock

Address: 1901 Keel Road

City: Ramsbottom, Illinois   Zip: 64736

Phone: 063 259 3730   Customer #: 18306

REMARKS:

10 yrs. Parts & Labor on the Piano
1 yr. Parts & Labor on Pianocorder

Delivery Address:

DESCRIPTION	PRICE	DISCOUNT	AMOUNT
New Samick 5'2" Grand Piano model G-1A			
# 820991 with Marantz P-101 # 11359			9500.00
		TOTAL AMOUNT	9500.00
		TRADE IN ALLOWANCE	2300.00
		SALES TAX	
		DEPOSIT	1000.00
		FINAL BALANCE	6200.00

PLEASE NOTE: All sales pending approval by management and verification of trade-in description.

If this contract is breached by the BUYER, the SELLER may take appropriate legal action, or, at its option, retain the deposit as liquidated damages.

Buyer's Signature

Figure 11.3 Sales contract.

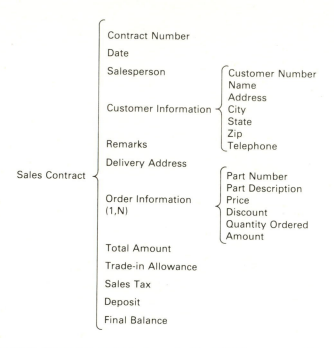

Figure 11.4 The sales contract shown in Fig. 11.3 is represented here by a Warnier–Orr diagram.

REPRESENTATION OF PROGRAM STRUCTURE

Figure 11.5 shows the same type of diagram representing a program structure. It shows the same program structure as the structure chart in Fig. 11.1. Warnier–Orr is more powerful than a structure chart because it can show the basic program control constructs: sequence, selection, and repetition.

When representing a program structure, each level in a Warnier–Orr diagram has three parts, BEGIN, process steps, and END. Each level is enclosed in vertical brackets, and the hierarchical structure is read from left to right. To indicate sequence in a Warnier–Orr diagram the processing steps are included at the same hierarchical level and are written in a vertical column one after another. A structure chart such as Fig. 11.1 does not show sequence.

The number of times a function is executed is indicated by a number or variable enclosed in parentheses below the name of the function. For example, in Fig. 11.5, the function PROCESS SUBSCRIPTION is executed S times, where S > 1. The (1,S) notation indicates a DO UNTIL structure. The termination condition for the structure is defined in footnote ?1. In Warnier–Orr diagrams, no control logic is included in the body of the diagram; instead, it is included in footnotes below the diagram.

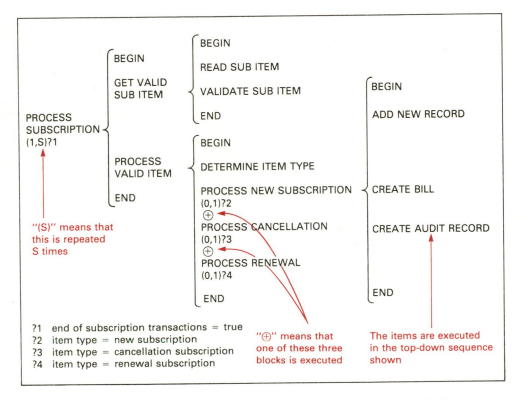

Figure 11.5   Warnier–Orr diagram for the Subscription System shown in the structure chart of Fig. 9.1, the HIPO diagram of Fig. 10.1, and Michael Jackson diagram of Fig. 12.3.

To indicate the selection construct, (0,1) is written below the function name. In Fig. 11.5, the functions PROCESS NEW SUB, PROCESS RENEWAL, and PROCESS CANCEL are each selectively performed. The inclusive or exclusive or, "+" or "⊕", is used in conjunction with (0,1) to indicate whether or not at most one alternative can be selected. Note that the ⊕'s separating PROCESS NEW SUB, PROCESS RENEWAL, and PROCESS CANCEL are used to indicate that a subscription can be of only *one* type.

However, the diagram body does *not* contain any information about what condition is tested to determine which function is selected. Footnotes ?2, ?3, and ?4 indicate the conditions that are tested to determine which alternative to execute. This is a weakness of the Warnier–Orr diagram. Incorporating the control logic into the body of the diagram would greatly improve its readability.

Box 11.1 summarizes the four basic constructs used to build Warnier–Orr diagrams.

BOX 11.1 Basic constructs of the Warnier–Orr diagram

Construct	Description	Data Example	Function Example
HIERARCHY name	A bracket is used to enclose the member of a set. Brackets are nested to show hierarchical levels. A bracket is always given a name. Brackets may be nested to show functional decomposition.	Employee file { Get valid sub item	{ Get valid sub item
SEQUENCE name { item 1 item 2 . . . item n	The sequence construct is used to show the ordering of members in a set by listing them one below another inside a bracket.	Employee { Employee name Employee address Employee number Employee sex Employee salary	Process New Subscription { BEGIN Add new record Create bill Create audit record END Note that BEGIN and END are used to delimit the sequence of steps in a function.
ALTERNATION (also called SEQUENCE or CASE) x (0,1) ⊕ y (0,1) ⊕ z (0,1)	The alternation construct is used to show partitioning into two or more mutually exclusive alternatives. The (0,1) below the name indicates that x occurs once or not at all. The ⊕ is the mutually exclusive or notation separating alternatives. It means that x, y, or z occurs—but only one of the three.	Employee pay type { Hourly (0,1) ⊕ Salaried (0,1)	Process Valid Subscription { Begin Process New Subscription (0,1) + Process Renewal (0,1) + Process Cancellation (0,1) END

REPETITION

The repetition structure is used to show that something occurs repeatedly. There are three forms:

(1) The first structure shows that the structure named L occurs n times. n may be a variable or a constant value.

(2) The second structure is a DO UNTIL structure that shows that function U occurs 1 to n times.

(3) The third structure is a DO WHILE structure that shows that function W occurs 0 to n times.

(Note that Warnier does *not* allow this form in his version of Warnier–Orr diagrams.)

Note: if U and W are data sets, then the notation means that the data set occurs 1 to n or 0 to n times.

$$L \atop (n)$$

$$U \atop (1,n)$$

$$W \atop (0,n)$$

Employee
(1,E)
{
Employee name
Employee address
Employee number
Employee type

Process
Subscription
(1,S)
{
BEGIN
Get valid sub item
Process valid item
END

143

CRITIQUE OF WARNIER–ORR DIAGRAMS

Warnier–Orr diagrams have been used extensively to design new systems and to document existing systems. Transferring a Warnier–Orr diagram into structured program code is usually quite simple because of its BEGIN-END block structure format.

Studies have shown that data structure documentation is more useful to an overall program understanding than is procedural documentation. Providing good data structure documentation is a major benefit of Warnier–Orr diagrams.

Warnier–Orr diagrams differ from HIPO diagrams in that they can be used to show the structure of both procedural and data components. They are similar to HIPO diagrams in that they both can be used to describe a system or program at varying degrees of detail during the functional decomposition process. Also, they both offer the advantage of making a program design more visible and understandable as it evolves. At a high level, HIPO diagrams and Warnier–Orr diagrams can give a clear representation of the structure of a system or program. However, when used at a low level, they both can become large and difficult to read.

Warnier–Orr diagrams are preferred over HIPO diagrams. They are a superior diagramming technique with one exception. Detail HIPO diagrams relate data to processing steps. Warnier–Orr diagrams do not have this valuable capability.

Warnier–Orr diagrams are an alternative to combination approaches such as structure charts and pseudocode or structure charts and Nassi–Shneiderman charts. They offer the advantage of using one technique for both high-level and detail design and for both procedure and data structure design. The major shortcoming is that Warnier–Orr diagrams do not show conditional logic as well as do other detail-level diagramming techniques. Another major problem is that they are not data-base oriented. They can only represent hierarchical data structures.

In general, Warnier–Orr diagrams are better suited for designing and documenting small rather than large problems, especially output-oriented problems with simple file structures.

REFERENCE

1. J. Warnier, *Logical Construction of Systems*, Van Nostrand Reinhold Company, New York, 1981, pp. 11–38.

12 MICHAEL JACKSON DIAGRAMS

INTRODUCTION   Warnier–Orr diagrams have the advantage that they represent both data structures and program structures. Michael Jackson techniques [1,2] have the same advantage. In addition, both Warnier–Orr and Jackson emphasize that the program structures should be derived from the data structures. The input data and output data of a program are used to create the program structure.

TREE STRUCTURE DIAGRAMS   Jackson views program structures and data structures as hierarchical structures. He uses a tree structure diagram to represent both.

The tree structure diagram looks very similar to the structure chart presented in Chapter 9 (see Fig. 9.1). Like the structure chart, it is composed of rectangular boxes arranged in levels and connected by lines. When used to show program structure, the tree structure diagram and the structure chart give the same information with the following exceptions:

- Only the structure chart shows the data passed between functional components.
- Only the tree structure diagram shows the control constructs of sequence, selection, and iteration.

A *tree structure diagram* is composed of four basic components: sequence, selection, iteration, and elementary.

Sequence

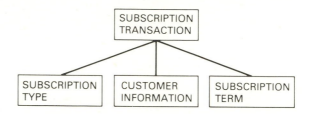

A *sequence component* is made up of a sequence of parts, occurring once each and in a specified order. The order of the parts is shown by reading the diagram from left to right. The example above is read: a SUBSCRIPTION TRANSACTION is composed of a SUBSCRIPTION TYPE followed by CUS-TOMER INFORMATION, followed by SUBSCRIPTION TERM. Notice that the sequence structures is a two-level structure. The first level names the component; the second level lists its parts.

Selection

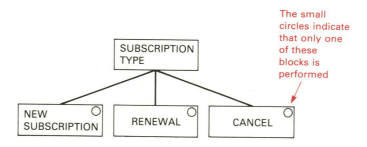

The small circles indicate that only one of these blocks is performed

A *selection component* is made up of two or more parts, exactly one of which occurs for each occurrence of the selection component. In the example above, a SUBSCRIPTION TYPE is either a NEW SUBSCRIPTION, a RE-NEWAL, or a CANCEL. The small circle in the upper right-hand corner indicates the parts of the selection component. Like the sequence component, the selection component is a two-level structure. The first level names the component; the second level lists the alternative parts.

Iteration

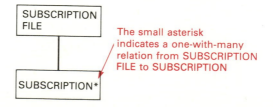

The small asterisk indicates a one-with-many relation from SUBSCRIPTION FILE to SUBSCRIPTION

The *iteration component* consists of zero, one, or more occurrences of its parts. In the example above, SUBSCRIPTION FILE consists of zero or more SUBSCRIPTIONS. The asterisk in the upper right-hand corner is used to indicate iteration. Again notice that the iteration structure is a two-level structure.

Elementary

```
┌──────────────┐
│   CHARGES    │
└──────────────┘
```

An elementary part is drawn in the tree structure diagram as one rectangular box. It corresponds to the lowest-level data item or program part in the design.

DATA STRUCTURE DIAGRAMS   Figure 12.1 shows a Jackson representation of data. This is the same employee file that is drawn with the Warnier–Orr technique in Fig. 11.2.

The diagram indicates that an EMPLOYEE FILE is composed of a FILE HEADER,

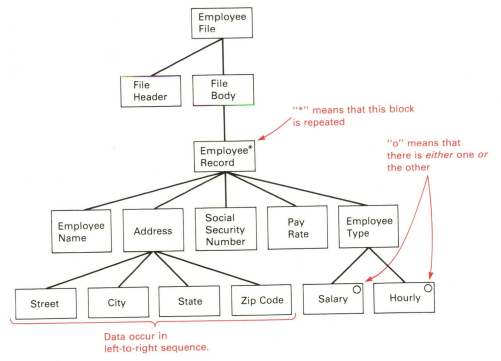

Figure 12.1   Jackson diagramming represents the data stream which enters or leaves a program by means of a hierarchy chart. These are the same data as those shown in the Warnier–Orr chart of Fig. 11.2.

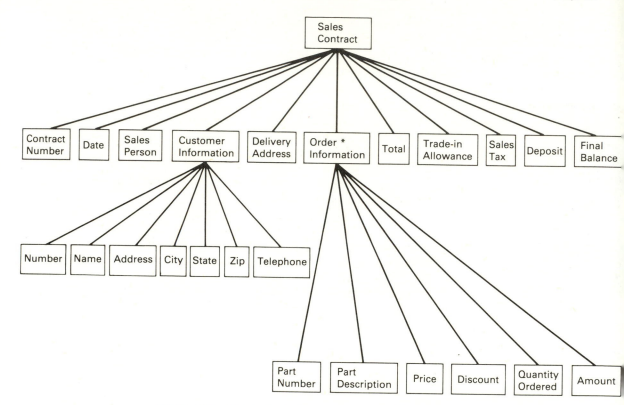

Figure 12.2   This is the Jackson tree structure diagram representing the sales contract shown in Fig. 11.3. It is equivalent to the Warnier–Orr diagram shown in Fig. 11.4.

then a FILE BODY. ADDRESS is composed of STREET, CITY, STATE, and ZIP CODE.

The asterisk in EMPLOYEE RECORD indicates that this occurs multiple times within FILE BODY. This is equivalent to the (1,E) indication on the Warnier–Orr chart. Warnier and Orr give somewhat more information than Jackson in that their chart can say *how many* times an item is repeated.

The ''o'' in the SALARY and HOURLY boxes indicates an exclusive/or situation. EMPLOYEE TYPE is either SALARY or HOURLY. This is equivalent to the ⊕ in the Warnier–Orr chart of Fig. 11.2.

Like Warnier–Orr, then, a Jackson chart shows sequence, selection, and iteration.

Figure 12.2 is a tree structure diagram of a sales contract shown in Fig. 11.3. Figure 11.4 is the equivalent Warnier–Orr diagram.

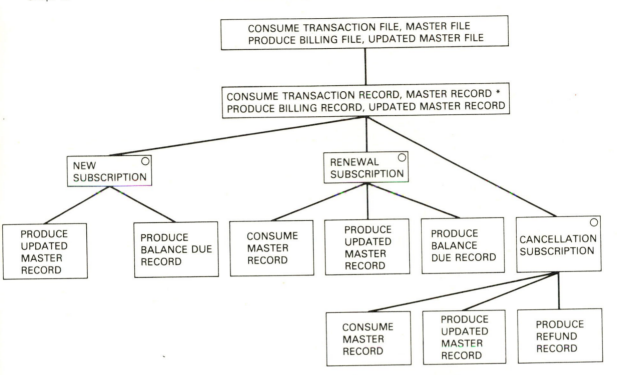

Figure 12.3   The Jackson tree structure diagram is used to represent a program structure or a data stream. This is the equivalent program structure that is drawn in the HIPO diagram of Fig. 10.1, the Warnier–Orr chart of Fig. 11.5, and the structure chart of Fig. 9.1.

PROGRAM STRUCTURE DIAGRAMS

Figure 12.3 shows the same type of diagram for representing the structure of programs. Again "o" in the top right-hand corner of the blocks means selection. Only one of the blocks NEW SUBSCRIPTION, RENEWAL SUBSCRIPTION, and CANCELLATION SUBSCRIPTION is executed. The diagram does not indicate how the choice is made. This is equivalent to the "⊕" in the Warnier–Orr charts.

Again "*" means repetition. The CONSUME RECORDS PRODUCE RECORDS block is repeated (which means repetition of everything that comprises this block, which is drawn underneath it). The diagram does not indicate what controls the repetition. This is similar to the "(1,S)" in Warnier–Orr charts.

The Jackson representation of data and programs is similar to the Warnier–Orr representation, but drawn vertically instead of horizontally. The reader should compare Figs. 12.1 and 12.3 with Figs. 11.2 and 11.5.

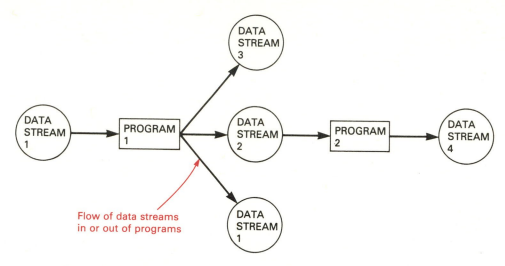

Figure 12.4   Jackson overview diagram, called a system network diagram. This gives similar information to that on a data flow diagram. The reader might compare it with Fig. 7.1.

SYSTEM NETWORK DIAGRAM

Jackson methodology (JSD, Jackson System Design) first designs the data structures which a program uses. It designs the input and output data streams. Then from these diagrams it designs the program structure.

First, an overall diagram is drawn showing the data streams that enter and leave the programs. Figure 12.4 illustrates this. Rectangles are used for programs and circles are used for data streams.

This diagram is similar to a data flow diagram, like that in Fig. 7.1, except that the notation is different and the rules for drawing a system network diagram are more formal than for a data flow diagram. A data flow diagram labels the arrows to show the data streams. Jackson refers to the diagram in Fig. 12.4 as a *system network diagram*.

The rules for drawing a system network diagram are:

- An arrow is used to connect a circle and a rectangle but is not used to connect two circles or two rectangles.

- Each circle may have at most one arrow pointing toward it and at most one arrow pointing away from it.

A data flow diagram could be used instead.

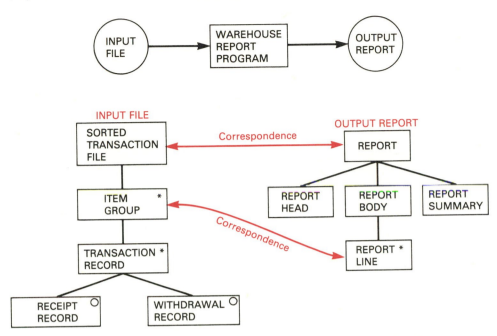

Figure 12.5 On the left is a Jackson drawing of an input file which is input to the WAREHOUSE REPORT PROGRAM. On the right is a drawing of the output report that is required from the program. The red arrows show one-to-one correspondence between input and output [3].

FROM DATA TO PROGRAMS

In Jackson's methodology, correct representations of input and output data streams lead to precise structures of programs. We can draw links between the input and output charts to show what data items correspond. Figure 12.5 illustrates this. It shows data structures which are the input and output of a WAREHOUSE REPORT PROGRAM. The connecting arrows show that there is a one-to-one correspondence between SORTED TRANSACTION FILE in the input and REPORT in the output, and between ITEM GROUP in the input and REPORT LINE in the output. In other words, one SORTED TRANSACTION FILE produces one REPORT. Similarly, one ITEM GROUP produces one REPORT LINE.

From these data structures a corresponding program structure is created which encompasses all parts of each data structure. This is shown in Fig. 12.6. Where there are one-to-one correspondences between the input and output, the program block says CONSUME input PRODUCE output. This block in the level below may be broken into more detail, showing computational operations or

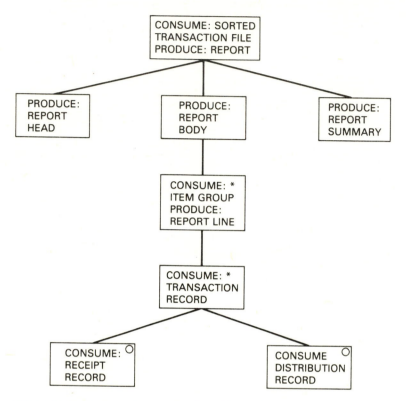

Figure 12.6   Program structure produced from the data structures (Michael Jackson methodology).

algorithms that link input and output. In Fig. 12.6 there is no separate block for CONSUME and PRODUCE as there would be on a structure chart. This is to emphasize that the data items consumed and produced correspond to one another.

Executable program operations can now be allocated to the program structure of Fig. 12.6, as shown in 12.7. Jackson methodology provides rules for checking both this allocation and the program structure.

Figure 12.8 shows Jackson's formally coded *structure text* derived from the structure of Fig. 12.7. The program operations written on Fig. 12.7 appear in the structure text of Fig. 12.8. Figure 12.9 shows the structure text converted into a program.

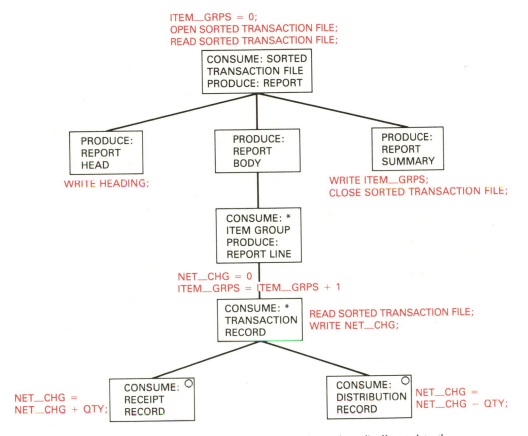

Figure 12.7 Executable program operations (shown in red) allocated to the program structure of Fig. 12.6 [3].

CRITIQUE OF JACKSON DIAGRAMS

Jackson's mapping of input data structures to output data structures is an aid to clear thinking in program design. Major users of the methodology claim that, at least for straightforward data processing, it leads to code with fewer problems than other forms of structuring in common use.

It does not help with complex program logic, and treats data-base systems as though they were essentially the same as file systems. The structure text of Fig. 12.8 is long-winded and not easy to read. It is longer than the program derived from it in Fig. 12.9, and much longer than the code needed with powerful fourth-generation languages.

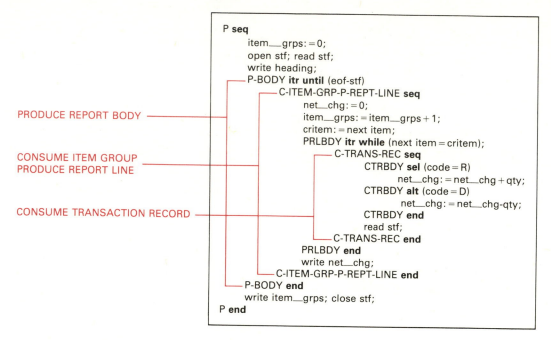

```
P seq
        item__grps: = 0;
        open stf; read stf;
        write heading;
      P-BODY itr until (eof-stf)
              C-ITEM-GRP-P-REPT-LINE seq
                   net__chg: = 0;
                   item__grps: = item__grps + 1;
                   critem: = next item;
                   PRLBDY itr while (next item = critem);
                        C-TRANS-REC seq
                             CTRBDY sel (code = R)
                                     net__chg: = net__chg + qty;
                             CTRBDY alt (code = D)
                                     net__chg: = net__chg-qty;
                             CTRBDY end
                             read stf;
                        C-TRANS-REC end
                   PRLBDY end
                   write net__chg;
              C-ITEM-GRP-P-REPT-LINE end
      P-BODY end
        write item__grps; close stf;
P end
```

PRODUCE REPORT BODY

CONSUME ITEM GROUP
PRODUCE REPORT LINE

CONSUME TRANSACTION RECORD

Figure 12.8   Jackson structure text derived from the program structure in Fig. 12.7 [3].

```
PB:        item__grps: = 0;
           open stf;
           read stf;
           write heading;
PBB:       do while (not eof-stf);
                net__chg: = 0;
                item__grps: = item__grps + 1;
                critem: = next item;
PRLBB:          do while (next item = critem);
CTRBB:               if (code = R) then
                          net__chg: = net chg + qty;
                     else if (code = D) then
CTRBE:                    net__chg: = net__chg-qty;
                          read stf;
PRLBE:          end;
                write net__chg;
PBE:       end;
           write summary;
PE:        close stf;
```

Figure 12.9   Final data structure design program.

154

REFERENCES

1. M. A. Jackson, *Principles of Program Design*, Academic Press, Inc., New York, 1975.

2. M. A. Jackson, *System Development*, Prentice-Hall, Inc., Englewood Cliffs, NJ, 1983.

3. This example is borrowed from one of the wittiest and best articles on structured techniques we have read: G. D. Bergland, "A Guided Tour of Program Design Methodologies," *IEEE Computing*, October 1981, pp. 13–37.

13 FLOWCHARTS

OVERVIEW STRUCTURE VERSUS DETAILED STRUCTURE

The diagrams that we have used so far do not show the detailed structure of programs; they show the overview structure or architecture. This and the following four chapters discuss techniques for showing program detail.

Techniques for showing overview structure and techniques for showing detail can be subdivided as shown in the following table:

Techniques for Showing Overview Structure of Programs	Techniques for Showing Logic Detail of Programs (Conditions, Case Structures, Loop Structures)
Functional decomposition Structure charts Data flow diagrams Warnier-Orr charts Michael Jackson charts HIPO diagrams (table of contents) Action diagrams HOS charts	Warnier–Orr charts HIPO (overview and detail) Action diagrams HOS charts Flowcharts Pseudocode or structured English Nassi–Shneiderman diagrams Decision trees Decision tables Finite-state diagrams

Note that Warnier–Orr diagrams, action diagrams, and HOS charts cover both the overview structure and the detailed logic.

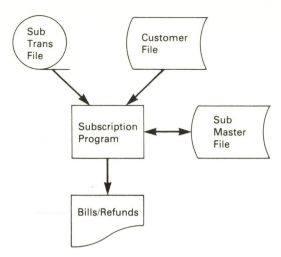

FLOWCHARTS

Flowcharts were one of the earliest forms of diagramming method. They were used by most analysts and programmers before the era of structured techniques. They are still often found in use today. Flowcharts can be replaced with smaller, cleaner, structured diagrams, and generally we do not recommend their use today. There are two types of flowcharts: system flowcharts and program flowcharts.

SYSTEM FLOWCHARTS

Figure 13.1 is an example of a system flowchart for the subscription system. It shows the basic input, output, and processing components for a system or program. Usually, input and output are represented as physical files. The processing components represent individual load units or job steps. The same information can be represented in an overview HIPO diagram (see Fig. 10.2). The systems flowchart serves as *operations* documentation, showing the computer operator how to execute the system.

PROGRAM FLOWCHARTS

Figure 13.2 is an example of a program flowchart for the module VALIDATE SUB ITEM in the subscription system. A program flowchart is primarily a coding tool. It shows in graphic form the sequence in which statements or process blocks will be executed and the control logic that governs their execution. Traditionally, program flowcharts have served two purposes. First, they have been used as a program design tool to plan detailed and complicated program logic. Second, they have been used as program documentation.

Because program flowcharts provide a sequential rather than a hierarchical representation of a program, they cannot clearly show program structure and the

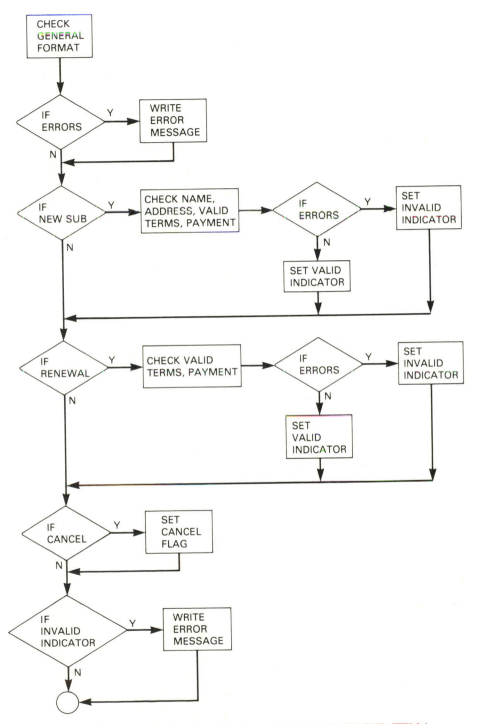

Figure 13.2 Program flowchart for the module VALIDATE SUB ITEM in the subscription system. A program flowchart is used to show detailed program logic.

interrelationships between procedural components. The lines with arrows drawn across flowcharts tend to encourage the use of undesirable GO TO statements. Hierarchical diagramming techniques such as those described in subsequent chapters are preferred over flowcharts to describe program structures and form a basis for computer-aided program design.

Some programmers draw hierarchical diagrams such as structure charts or Warnier–Orr diagrams but then draw a flowchart to represent the internal logic of a program module. It is desirable to have *one* diagramming technique that is easy to use both for the overall program structure and for the internal logic.

Some programmers prefer pseudocode to flowcharts. It is a more compact way to describe detailed program logic. A flowchart can easily be three or four times larger than a comparable pseudocode representation. Pseudocode has the disadvantage that it is not graphic.

Figure 14.1 shows the pseudocode version of the logic in the flowchart of Fig. 13.2. Figure 15.1 shows a Nassi–Shneiderman version of Fig. 13.2. Figure 16.3 shows an action diagram version of Fig. 13.2.

In the case of unstructured programs, flowcharts offer the only graphic technique for describing program form and logic. Experiments have shown that flowcharts showing control flow information made debugging easier than when source code is used by itself [1].

FLOWCHART SYMBOLS

Figure 13.3 shows the flowchart symbols in conventional use.

CRITIQUE OF FLOWCHARTS

Generally, the flowchart is *not* considered to be a structured diagramming technique. Its utility is limited to small programs (fewer than 10,000 lines). For larger programs, flowcharts become very cumbersome to use. But even in the case of smaller programs, they normally should not be used as a program design tool. Flowcharts encourage a view of the problem that is likely to lead to a poorly structured program.

At a high level, flowcharts, as commonly drawn, tend to encourage a physical view of the system. The flowcharts for on-line and off-line systems which carry out the same logical functions, for example, are very different. An overall drawing of the functions is needed which leaves a designer free to decide later what should be on-line or off-line. Flowcharts thus tend to encourage a physical view of the system before the overall logical requirements are understood. Gane and Sarson refer to this as being "prematurely physical" [2].

Flowcharts can show detailed logic (in an unstructured fashion) but do not give a useful overview of the system functions in the way that data flow diagrams and functional decomposition charts do. In particular, detailed flowcharts should be avoided as a form of program documentation. Experiments by Shneiderman,

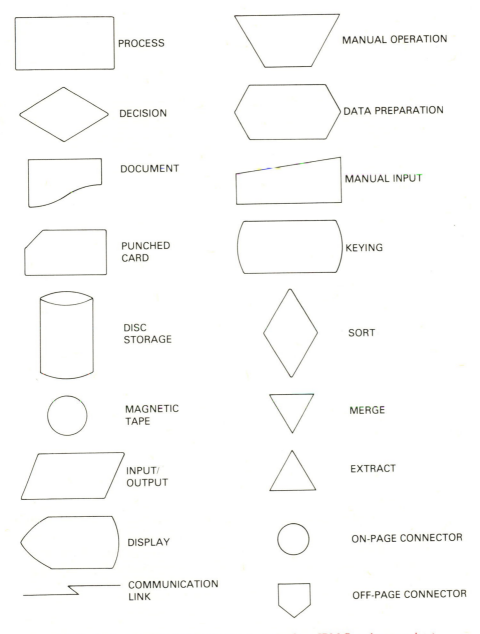

PROCESS

MANUAL OPERATION

DECISION

DATA PREPARATION

DOCUMENT

MANUAL INPUT

PUNCHED CARD

KEYING

DISC STORAGE

SORT

MAGNETIC TAPE

MERGE

INPUT/ OUTPUT

EXTRACT

DISPLAY

ON-PAGE CONNECTOR

COMMUNICATION LINK

OFF-PAGE CONNECTOR

Figure 13.3   Conventional flowchart symbols (from IBM flowchart template).

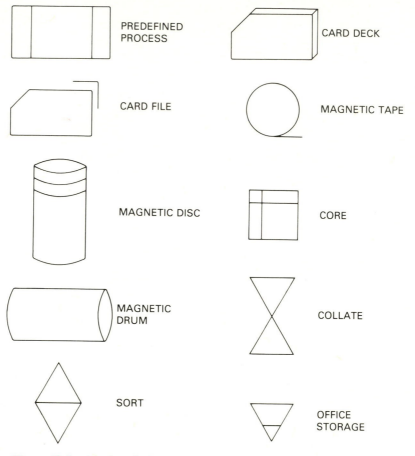

Figure 13.3 *(continued)* Composite symbols (from IBM flowcharting template).

for example, could not demonstrate that detailed flowcharts aided program understanding, debugging, or modification [3]. When a well-structured program listing is available, pseudocode or graphic documentation (e.g., Nassi–Shneiderman chart) is more useful. A detailed flowchart is often larger than the program it represents even though it does not contain declarations or data structure information.

Useful documentation should be at a higher level than the actual code. Shneiderman comments that to be effective it should probably not be greater than one-tenth of the code length (although this usually does not apply with fourth-generation languages) [4]. Also, it should use problem-related rather than program-related terminology. Flowcharts fail to meet any of these criteria.

REFERENCES

1. J. Brooke and K. Duncan, ''An Experimental Study of Flowcharts as an Aid to Identification of Procedural Faults,'' *Ergonomics*, Vol. 23, No. 4 (1980): 387–399.

2. C. Gane and T. Sarson, *Structured Systems Analysis: Tools and Techniques*, IST Databooks, New York, 1977.

3. B. Shneiderman, R. Mayer, D. McKay, and P. Heller, ''Experimental Investigations of the Utility of Detailed Flowcharts in Programming,'' *CACM*, Vol. 20 (1977): 373–381.

4. B. Shneiderman, ''Control Flow and Data Structure Documentation,'' *CACM*, Vol. 25, No. 1 (January 1982): 55–63.

14 STRUCTURED ENGLISH AND PSEUDOCODE

INTRODUCTION
As noted in Chapter 13, flowcharts do not represent structured design. At a high level they encourage being "prematurely physical." At a detailed level they encourage GO TOs and non-structured code. They were boycotted by the structured enthusiasts with the comment that they do more harm than good.

At first there was not a diagramming technique to replace flowchart representation of program detail. Flowcharts were replaced with structured English and pseudocode. A typical structured methodology draws tree charts like Figs. 8.4, 9.1, 11.3, and 12.2 in which each module is a black box whose contents are not shown on the chart. The contents are represented separately with structured English or pseudocode.

Even at the level of structured English we believe a picture is worth a thousand words. Analysts and users make mistakes with structured English because it does not reveal the logic structure with the same clarity as a good diagram.

WHY SHOULD ENGLISH BE STRUCTURED?
Typical English narrative describing specifications can contain all manner of ambiguities. The language of lawyers is supposed to be highly professional and precise, but in fact it is common to find ambiguities in contracts. Lawyers achieve what precision they do have at the expense of lengthy language which is often misinterpreted because it is tedious:

> The service fee shall be computed by adding the revenue accruing from sales of tapes on a monthly basis to the monthly rental revenue for tapes and multiplying by twelve percent except in the case where the revenue accruing from the sales of tapes exceeds $US 5000 in any one calendar month in which case the service fee shall be computed by adding the revenue accruing from

the sales of tapes on a monthly basis multiplied by fifteen percent to the monthly rental revenue for tapes multiplied by twelve percent; notwithstanding the above in the circumstance that the revenue accruing from sales of tapes exceeds $US 8000 in any one calendar month the service fee shall be computed by adding the revenue accruing from the sales of tapes on a monthly basis multiplied by eighteen percent to the monthly rental revenue for tapes multiplied by twelve percent.

In structured English we would say:

```
IF monthly-sales-revenue > 8000
    Fee = monthly-sales-revenue * .18 + monthly-rental-revenue * .12
IF monthly-sales-revenue > 5000
    Fee = monthly-sales-revenue * .15 + monthly-rental-revenue * .12
ELSE
    Fee = monthly-sales-revenue * .12 + monthly-rental-revenue * .12
```

This is easier to read and leads more directly to the program code.

The example above is simple and the lawyer's wording is precise. Many DP situations are much more complex, with multiple nested conditions and repetitions. There is a greater need for clarity of structure.

AMBIGUITIES

In casual English many ambiguities arise. Consider the following clause:

all customers with more than $5000 in their account who have an average monthly balance exceeding $500 or who have been customers for more than five years

There are two possible meanings to this:

1. "all customers (with more than $5000 in their account AND an average monthly balance exceeding $500) OR who have been customers for more than five years"
2. "all customers with more than $5000 in their account AND (an average monthly balance exceeding $500 OR who have been customers for more than five years)"

The parentheses give the clause more structure and remove the ambiguity. In general, when a sentence contains "and" and "or" words together, there may be ambiguity which can be clarified with parentheses.

"If A and B or C" should become "if A and (B or C)" or "if (A and B) or C."

There is another ambiguity in the clause above: ''average monthly balance.'' We need to know over how many months the average is taken.

STRUCTURED ENGLISH

Figures 14.1 and 14.2 show typical examples of structured English. The figures have several important properties:

- They are written in such a way that a user could understand them.

- They are hierarchically structured and use indentation to reveal this structure.

- They have a similar structure to the program code that will be used to implement them.

- Comments that will not be translated into program code are marked with asterisks.

```
VALIDATE SUB ITEM

   * VALIDATE GENERAL FORMAT.
     CHECK GENERAL FORMAT.
     IF ERRORS
         WRITE ERROR-MESSAGE
     ENDIF.

   * VALIDATE SPECIAL FORMAT.
     IF NEW SUB
         CHECK NAME AND ADDRESS
         CHECK FOR NUMERIC ZIP
         CHECK FOR VALID TERMS
         CHECK FOR PAYMENT
         IF ERROR
             SET INVALID INDICATOR
         ELSE
             SET VALID INDICATOR
         ENDIF
     ENDIF.
     IF RENEWAL
         CHECK FOR VALID TERMS
         CHECK FOR PAYMENT
         IF ERROR
             SET INVALID INDICATOR
         ELSE
             SET VALID INDICATOR
         ENDIF
     ENDIF.
     IF CANCEL
         SET CANCEL FLAG
     ENDIF.
     IF INVALID INDICATOR
         WRITE ERROR MESSAGE
     ENDIF.
```

Figure 14.1   Structured English is a narrative notation used to define procedural logic. This example describes a structured version of the logic shown in the flowchart of Fig. 13.2.

```
ORDER ENTRY:
        OBTAIN CUSTOMER ORDER DETAILS
            IF (CUSTOMER IS VALID)
            THEN SET UP CUSTOMER DETAILS FOR ORDER HEADER RECORD
                    SET UP ORDER DELIVERY ADDRESS FOR ORDER HEADER RECORD
                    WRITE ORDER HEADER RECORD
            ELSE (CUSTOMER IS NOT VALID)
                    ISSUE MESSAGE "INVALID CUSTOMER"
                    EXIT ORDER ENTRY

        REPEAT UNTIL (ORDER IS COMPLETE)
            OBTAIN ORDER LINE ITEM DETAILS
            IF (ORDERED PRODUCT IS VALID)
                AND IF (QUANTITY ORDERED IS AVAILABLE)
                    THEN SET UP PRODUCT DETAILS FOR LINE ITEM RECORD
                            SET UP QUANTITY ORDERED FOR LINE ITEM RECORD
                            DECREMENT PRODUCT QUANTITY ON HAND BY QUANTITY ORDERED
                            UPDATE PRODUCT RECORD
                            WRITE LINE ITEM RECORD
                    ELSE (QUANTITY ORDERED IS NOT AVAILABLE)
                            SET UP PRODUCT DETAILS FOR BACKORDER RECORD
                            SET UP QUANTITY BACKORDERED FOR BACKORDER RECORD
                            WRITE BACKORDER RECORD
                ELSE (ORDERED PRODUCT IS NOT VALID)
                    ISSUE MESSAGE "INVALID PRODUCT"
        END REPEAT
        PREPARE PACKING SLIP
        WRITE PACKING SLIP
    EXIT
```

Figure 14.2   Structured English example of typical order entry logic.

FOUR BASIC TYPES OF STRUCTURE

Structured English, like any other means of representing program structures, should be designed to show four basic constructs:

1. *Sequence*: Simple, top-to-bottom sequence is used.

2. *Condition*: If a certain condition applies, then a given action will be taken; if not, a different given action may be specified.

3. *Case*: One of several possible cases exists. The structure shows what action is taken for each possible case. A mutually exclusive set of conditions is a *case* structure.

4. *Repetition*: A given set of operations is repeated. The condition that terminates this repetition is shown. Repetition structures are of two types:

   - REPEAT WHILE. The operations are repeated while a specified condition applies. This condition is tested *before* the execution of the operations.

   - REPEAT UNTIL. The operations are repeated until a specified condition exists. This condition is tested *after* the execution of the operations.

Keywords should be used for these structures in structured English. These words should themselves be English so that they are easily understandable by end users. Typical words are as follows:

Sequence

No keyword is necessary to show sequence. The sequence may be preceded by a title. The end of the sequence may be indicated with the word EXIT possibly followed by the sequence title.

Condition

IF and ELSE are commonly used. Sometimes IF . . . THEN . . . is used. THEN is not really necessary; a new line shows what would follow the THEN.

To make clear where an IF clause ends, it is valuable to write ENDIF.

Case

IF . . . ELSEIF . . . ELSEIF . . . ELSE is sometimes used for mutually exclusive conditions. Sometimes the case construct has its own words, different from an IF: for example, SELECT . . . WHEN . . . WHEN . . . WHEN.

To make clear where a case clause ends, it is valuable to write ENDIF, ENDSELECT, and so on.

Repetition

DO WHILE, REPEAT WHILE, or LOOP WHILE are used for repeat-while repetition; DO UNTIL, REPEAT UNTIL, or LOOP UNTIL are used for repeat-until repetition.

FOR ALL may be used to show that all items in a given set are to be processed: for example, "FOR ALL CUSTOMER_ORDER RECORDS." FOR EACH may be used to show that *each* item in a given set that meets certain criteria is processed. FOR EACH may be qualified with WHERE: for example, "FOR EACH PART WHERE QUANTITY_ON_HAND > 5000."

To make clear which is the end of the block that is to be repeated, it is necessary to write END, ENDDO, ENDREPEAT, ENDLOOP, or ENDFOR.

In addition to the foregoing keywords, certain keywords may be used to express logic: for example, AND, OR, GT (greater than), LT (less than), GE (greater than or equal), and LE (less than or equal).

An installation should select a given set of keywords for structured English and make these an installation standard.

In writing structured English it is useful to capitalize the keywords, names of program blocks, and names of items in the data dictionary. Everything else should be noncapitalized.

KEYWORDS FROM FOURTH-GENERATION LANGUAGES

By now, every efficient data processing installation ought to have selected at least one fourth-generation language [1]. Many fourth-generation languages contain keywords like those above. If such a language is intended to have major use in an installation, the keywords of the language may be selected as the installation standard for structured English. This has the advantage that the structured English is easier to translate into code and it may help and encourage the end users to employ the language themselves [2].

BOX 14.1 Rules for writing structured English

- The structures are indented to show the logical hierarchy.
- *Sequence*, *condition*, *case*, and *repetition* structures are made clear.
- The sequence structure is a list of items where each item is placed on a separate line. If the item requires more than one line, continuation lines are indented. The end of an item is punctuated with a semicolon (;) (see Fig. 14.3).
- Keywords are used to make the structures clear: for example, IF, THEN, ELSE, ENDIF, REPEAT WHILE, REPEAT UNTIL, END__REPEAT, EXIT.
- Keywords are used for logic: AND, OR, GT (greater than), LT (less than), GE (greater than or equal to), LE (less than or equal to).
- The choice of keywords should be an installation standard.
- The keyword set may be selected to conform to a fourth-generation language. (They do, however, remain language-independent descriptions.)
- Blocks of instructions are grouped together and given a meaningful name which describes their function.
- Keywords and names that are in the data dictionary are capitalized; names of program blocks are capitalized; other words are noncapitalized.
- Comments lines are delimited with a beginning asterisk and a terminating semicolon.
- Parentheses are used to avoid AND/OR and other ambiguities.
- End words such as ENDIF, ENDREPEAT, and EXIT are used to make clear where the structure ends.
- Within the foregoing constraints the wording should be chosen to be as easy as possible for end users to understand.

This concept offends purists who believe that structured English should be independent of the programming language. In practice, however, tailoring the specifications to a fourth-generation language *does* help, and it is trivial for a coder to code them in a different language. All programming languages have condition, case, and repetition constructs which are broadly similar, so select the words of one such language as the set that the installation uses.

**RULES FOR
STRUCTURED
ENGLISH**

Box 14.1 gives a set of rules which should govern the use of structured English. Figure 14.3 gives a version of Fig. 14.2 that follows these rules.

```
ORDER_ENTRY:

        FOR ALL orders
            Obtain CUSTOMER record;
            IF CUSTOMER# is valid
                Set up customer details for ORDER_HEADER record;
                Set up order delivery address for ORDER_HEADER record;
                Write ORDER_HEADER record;
            ELSE
                Issue "invalid customer" message;
                QUIT ORDER_ENTRY;
            ENDIF;
            FOR ALL items ordered
                Obtain PRODUCT record;
                IF PRODUCT# is valid
                    IF quantity ordered is available
                        Set up product details for LINE_ITEM record;
                        Set up quantity ordered for LINE_ITEM record;
                        Decrease product quantity on-hand by
                          quantity ordered;
                        Update PRODUCT record;
                        Write LINE_ITEM record;
                    ELSE
                        Set up product details for BACKORDER record;
                        Set up quantity backordered for BACKORDER records;
                        Write BACKORDER record;
                    ENDIF;
                ELSE
                    Issue "invalid product" message;
                ENDIF.
            ENDFOR.
            Prepare package slip;
            Print package slip;
        ENDFOR;
EXIT ORDER_ENTRY.
```

Figure 14.3   Figure 14.2 rewritten with the rules of Box 14.1. The keywords used here are those of the fourth-generation language IDEAL.

```
P seq
    item__grps: = 0;
    open stf; read stf;
    write heading;
    P__BODY itr until (eof-stf)
            C-ITEM-GRP-P-REPT-LINE seq
                    net__chg: = 0;
                    item__grps: = item__grps + 1;
                    critem: = next item;
                    PRLBDY itr while (next item = critem);
                            C-TRANS-REC seq
                                    CTRBDY sel (code = R)
                                            net__chg: = net chg + qty;
                                    CTRBDY alt (code  =  D)
                                            net__chg: = net__chg-qty;
                                    CTRBDY end
                                    read stf;
                            C-TRANS-REC end
                    PRLBDY end
                    write net chg;
            C-ITEM-GRP-P-REPT-LINE end
    P-BODY end
    write item__grps; close stf;
P end
```

Figure 14.4   Highly cryptic form of pseudocode that cannot be read or checked by end users [3].

PSEUDOCODE

We may distinguish between structured English and pseudocode (although sometimes these words are used interchangeably). Pseudocode uses more formal notation, more oriented to the DP professional, whereas structured English is designed so that end users can read it after minimal training.

In practice one can observe a spectrum ranging from entirely informal structured English without keywords, to a pseudocode notation which is close to the outline of the final program and which is difficult for end users to read. Some pseudocode instructions can be translated one for one into similar-looking program code instructions; some are translated into many program instructions.

Figure 14.4 gives an example of pseudocode which most end users would never dare to examine. In this sense it is very different from Figs. 14.1 to 14.3.

STRUCTURED TEXT OF MICHAEL JACKSON

Figure 14.4 uses Michael Jackson's variant of pseudocode [3]. It shows pseudocode for the program structure shown in Fig. 12.6.

Michael Jackson employs a formal notation for pseudocode, which he calls *structure text*. It accompanies a program structure chart [or as Jackson defines it, a tree structure diagram (Chapter 12)] and is used to complete the program design by defining the control logic. Figure 14.5 shows an example of structure text for the module VALIDATE SUB ITEM in the

```
VALIDATE-SUB-ITEM seq
        process CHECK-GENERAL-FORMAT;
        ERROR-CHECK select (ERROR-SW = 'ON')
                write ERROR-MESSAGE
        ERROR-CHECK end;
        VALIDATE-SPECIAL-FORMAT select (SUB-TYPE = 'NEW')
                VALIDATE-NEW-SUB seq
                        check name and address;
                        check numeric zip;
                        process CHECK-TERMS;
                        check payment;
                        ERROR-CHECK select (ERROR-SW = 'ON')
                                set INVALID-INDICATOR
                        ERROR-CHECK alt (ERROR-SW = 'OFF')
                                set VALID-INDICATOR
                        ERROR-CHECK end;
                VALIDATE-NEW-SUB end;
        VALIDATE-SPECIAL-FORMAT alt (SUB-TYPE = 'RENEWAL')
                VALIDATE-RENEWAL-SUB seq
                        process CHECK-TERMS;
                        check payment;
                        ERROR-CHECK select (ERROR-SW = 'ON')
                                set INVALID-INDICATOR
                        ERROR-CHECK alt (ERROR-SW = 'OFF')
                                set VALID-INDICATOR
                        ERROR-CHECK end;
                VALIDATE-RENEWAL-SUB end;
        VALIDATE-SPECIAL-FORMAT alt (SUB-TYPE = 'CANCEL')
                set CANCEL-FLAG
        VALIDATE-SPECIAL-FORMAT end;
        ERROR-CHECK select (INVALID-INDICATOR = 'ON')
                write ERROR-MESSAGES
        ERROR-CHECK end;
VALIDATE-SUB-ITEM end.
```

Figure 14.5   The pseudocode shown in Fig. 14.1 is rewritten using the Jackson structure text notation, which is more formal.

subscription system. Structured English for the same structure is shown in Fig. 14.1.

The rules for writing the control constructs of sequence, selection, and iteration are given below.

Sequence

```
N seq
    part-1;
    part-2;
    part-3;

        .
        .
        .

    part-n;
N end
```

N is the name of the sequence construct. In Fig. 14.5 VALIDATE-SUB-ITEM is the name of the first sequence construct listed. The parts are a list of one or more programming-level statements, high-level instructions, or control constructs. The parts are separated by a semicolon (;). The entry point to the sequence construct is N **seq** and the exit point is N **end**. The indentation shown above is used to clarify the structure. Parts are executed in the order in which they are listed.

Selection

```
N select (condition-1)
      part-1;
N alt (condition-2)
      part-2;
         .
         .
         .
N alt (condition-m)
      part-m;
N end
```

In Fig. 14.5, VALIDATE-SPECIAL-FORMAT is a selection construct. It has three alternative parts: VALIDATE-NEW-SUB **seq**, VALIDATE-RE-NEWAL-SUB **seq**, and set CANCEL-FLAG. One of these three alternative parts is executed, depending on the value of SUB-TYPE.

The select construct may have two or more parts. Only one part may be executed at one time. The entry point of the select construct is the name (N) followed by **select;** and the exit point is the name (N) followed by **end.** The indentation shown above is suggested to clarify the structure.

Iteration

```
N iter while (condition)
      part;
N end
```

The iteration construct is used to represent a loop. Its part is executed repeatedly while the condition is true. The iteration construct has an entry point composed of its name (N) followed by **iter** and an exit point composed of its name (N) followed by **end.**

Figure 14.4 is similar to Fig. 14.5 but is condensed by the use of mnemonics.

CRITIQUE OF STRUCTURED ENGLISH

Structured English has been a useful tool for describing program logic. We believe that tools which are more diagrammatic are better, as shown in the following chapters.

```
Three fields are defined, called "RECEIVED," "DISTRIBUTED,"
and "NETCHANGE," as part of the file called "WAREHOUSE."


DEFINE FILE WAREHOUSE
RECEIVED/19 = IF CODE IS 'R' THEN QUANTITY ELSE 0;
DISTRIBUTED/19 = IF CODE IS 'D' THEN QUANTITY ELSE 0;
NET_CHANGE/19 = RECEIVED-DISTRIBUTED;
END


The report is then produced as follows:


TABLE
FILE WAREHOUSE
SUM RECEIVED AND DISTRIBUTED AND NET_CHANGE
BY PRODUCT
END
```

Figure 14.6 Executable program code written with the fourth-generation language RAMIS II for the problem described with pseudocode in Fig. 14.4 and illustrated in Figs. 12.6 to 12.7. The executable program code is simpler than the pseudocode!

Figure 14.3 represents the form of structured English that we perceive to be generally the most useful. Figure 14.4 represents a form of pseudocode that we perceive as insufficiently user-friendly. It is difficult to read and even more difficult to write. A nontechnical end user and most ordinary data processing analysts would give up trying to decipher its meaning after the first few lines. If program names are long, the structure text can become a mess and invites mistakes. If program names are not descriptive, it becomes totally unreadable.

It is interesting to reflect that the *executable* program code of some fourth-generation languages is simpler and easier to read than some of the more cryptic breeds of pseudocode. In RAMIS II, for example, the problem pseudocoded in Fig. 14.4 can be executed simply with the code shown in Fig. 14.6.

REFERENCES

1. James Martin, *Fourth-Generation Languages*, Savant Institute, Carnforth, Lancashire, UK, 1983.

2. James Martin, *Application Development Without Programmers*, Prentice-Hall, Inc., Englewood Cliffs, NJ, 1982.

3. G. D. Bergland, "A Guided Tour of Program Design Methodologies," *IEEE Computing*, October 1981, pp. 13–37.

15 NASSI–SHNEIDERMAN CHARTS

INTRODUCTION Flowcharts, as we noted in Chapter 13, are not a good technique for describing structured programs. They give an unstructured view of the world and tend to lead to programs with GO TOs rather than hierarchically structured programs. Nevertheless, psychologists' experiments have confirmed what we might think is blatantly obvious—that clear diagrams do help in creating programs, and also in debugging them [1].

I. Nassi and B. Shneiderman set out to replace the traditional flowchart with a chart that offers a structured, hierarchical view of program logic [2]. Nassi–Shneiderman charts are used for detailed program design and documentation. N. Chapin uses a broadly similar type of diagram which he calls *Chapin charts*.

Nassi–Shneiderman charts (N-S charts) represent program structures that have one entry point and one exit point and are composed of the control constructs of sequence, selection, and repetition. Whereas it is difficult to show nesting and recursion with a traditional flowchart, it is easy with an N-S chart. Also, it is easy to convert an N-S chart to structured code.

An N-S chart consists of a rectangular box representing the logic of a program module. It is often used to represent what is in one of the blocks on a structure chart like that of Fig. 9.1. Figure 15.1 shows the logic inside the block VALIDATE SUB ITEM of Fig. 9.1.

The box of the N-S chart is intended to be drawn on one page. It should therefore not have more than about 15 to 20 subdivisions. When an N-S chart becomes too large, subfunctions are separated out and drawn on another N-S chart. It was designed for use with top-down, stepwise refinement methods.

Various structures are nested inside the box of an N-S chart to show the process logic. A branch instruction is *not* permitted.

VALIDATE SUB ITEM:

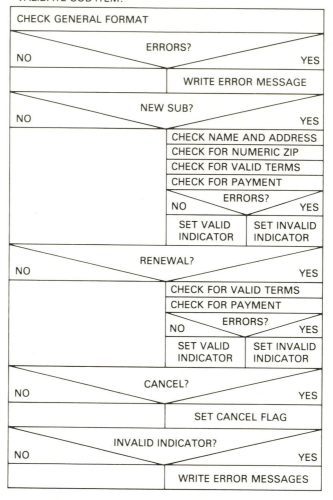

Figure 15.1   A Nassi–Shneiderman chart is used to show the detailed design for a program module. This chart shows the logic inside the VALIDATED SUB ITEM block of Fig. 9.1.

CONTROL CONSTRUCTS

Each basic control construct used in structured programming can be represented by a N-S chart symbol.

Sequence

Sequence is shown by a vertical stack of process boxes:

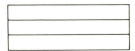

For example, Fig. 15.1 describes the sequence of four processes: CHECK NAME AND ADDRESS, CHECK FOR NUMERIC ZIP, CHECK FOR VALID TERMS, and CHECK FOR PAYMENT.

Selection

Selection (IF-THEN-ELSE) is shown by dividing the process box into five parts. The top half is divided into three triangles. The topmost triangle contains the condition to be tested. The bottom triangles indicate the ''true'' part and the ''false'' part of the IF-THEN-ELSE. The bottom half is divided into the ''true'' process box and the ''false'' process box, p-1 and p-2, respectively.

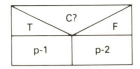

For example, in Fig. 15.1, the last condition test in the chart is to check the INVALID-INDICATOR. If the test is true, then the process WRITE ERROR MESSAGES is performed. If the test is false, nothing is done since the ''false'' process is null.

Note that the selection structure can be nested. In Fig. 15.1, the condition test for ERRORS is nested within the condition test for RENEWAL.

Case

The condition structure can be extended to the CASE structure in which a selection is made from multiple mutually exclusive choices, as follows:

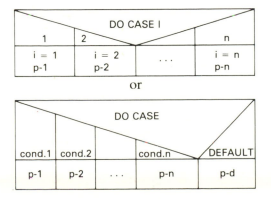

VALIDATE SUB ITEM:

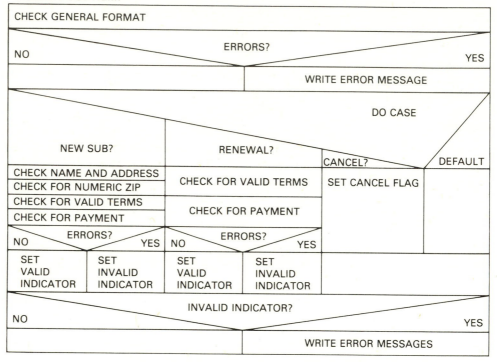

Figure 15.2   The function VALIDATE SUB ITEM from the subscription system can be implemented using the CASE structure.

The N-S chart shown in Fig. 15.1 could alternatively have been designed using the CASE structure. This is shown in Fig. 15.2.

Repetition

Repetition is indicated by a DOWHILE or a DOUNTIL structure.

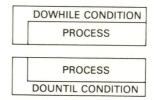

Notice that in the DOWHILE structure the condition is tested first, and then if the condition is true, the process is performed; but in the DOUNTIL structure the process is performed first, and then the condition is tested.

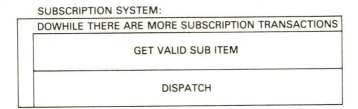

Figure 15.3   The function SUBSCRIPTION SYSTEM is performed repetitively until there are no more subscription transactions to process.

In the subscription system a repetition structure is used to indicate that the subscription process of validating and processing a subscription is performed once for each subscription on the transaction file. This is shown in Fig. 15.3.

CRITIQUE OF THE NASSI–SHNEIDERMAN CHART

The Nassi–Shneiderman chart is a diagramming technique used primarily for detail program design. It is a poor tool for showing the high-level hierarchical control structure for a program. Although it can show the basic procedural components (as shown in Fig. 15.3), it does not show the interfaces connecting these components.

It is an alternative to traditional program flowcharts, detail HIPO diagrams, and pseudocode. Among these techniques, the N-S chart is the easiest to read and the easiest to convert to program code. However, it is only a procedure design tool and cannot be used to design data structures. In addition, although it is easy to read, it is not always easy to draw. It can take three to four times longer to draw an N-S chart than to write the equivalent pseudocode. Once written, however, the N-S chart is easier to read than pseudocode.

Another major shortcoming of the N-S chart is that it is not data-base oriented. It does not link to a data model or to a data dictionary.

REFERENCES

1. J. Brook and K. Duncan, ''An Experimental Study of Flowcharts as an Aid to Identification of Procedural Faults,'' *Ergonomics*, Vol. 23, No. 4 (1980): 387–399.

2. I. Nassi and B. Shneiderman, ''Flowchart Techniques for Structured Programming,'' *ACM SIGPLAN Notices*, Vol. 8, No. 8 (August 1973): 12–26.

16 ACTION DIAGRAMS

Of the diagramming techniques we have described, some are usable for the *overview* of program structure and some are usable for the *detailed* program logic. Structure charts, HIPO diagrams, Warnier–Orr diagrams, and Michael Jackson charts draw overall program structures, but not the detailed tests, conditions, and logic. Their advocates usually resort to structured English or pseudocode to represent the detail. Flowcharts and Nassi–Shneiderman charts show the detailed logic of a program, but not the structural overview.

There is no reason why the diagramming of the *overview* should be incompatible with the diagramming of the *detail*. Indeed, it is highly desirable that these two aspects of program design should employ the same type of diagram because complex systems are created by successively filling in detail (top-down design) and linking together blocks of low-level detail (bottom-up design). The design needs to move naturally between the high levels and low levels of design. The low level should be a natural extension of the high level. *Action diagrams* achieve this. They give a natural way to draw program overviews such as structure charts, HIPO, or Warnier–Orr diagrams, *and* detailed logic such as flowcharts or Nassi–Shneiderman charts. They were originally designed to be as easy to teach to end users as possible and to assist end users in applying fourth-generation languages.

Glancing ahead, Figs. 16.2 and 16.3 show simple examples of action diagrams. Figure 16.10 shows an extension of Fig. 16.2.

BRACKETS   A program module is drawn as a bracket, thus:

Brackets are the basic building blocks of action diagrams. The bracket can be of any length, so there is space in it for as much text or detail as is needed.

Inside the bracket is a sequence of operations. A simple control rule applies to the bracket. You enter it at the top, do the things in it in a top-to-bottom sequence, and exit at the bottom.

Inside the bracket there may be other brackets. Many brackets may be nested. The nesting shows the hierarchical structure of a program. Figure 16.1 shows the representation of a hierarchical structure with brackets.

Some brackets are *repetition* brackets. The items in the bracket are executed multiple times. The repetition bracket has a double line at its top, thus:

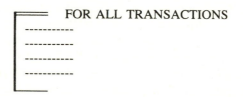

When one of several processes is to be used (mutually exclusive selection), a bracket with multiple divisions is used:

```
 ┌─── PROCESS NEW SUBSCRIPTION
 │
 ├─── PROCESS RENEWAL
 │
 ├─── PROCESS CANCELLATION
 └───
```

This is the programmer's CASE structure. One, and only one, of the divisions in the bracket above is executed. This replaces the "⊕" of Warnier–Orr, or the "o" of Michael Jackson, as shown below:

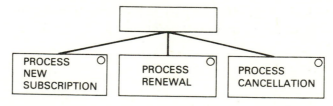

DECOMPOSITION DIAGRAM,
STRUCTURE CHART, HIPO CHART, ETC.

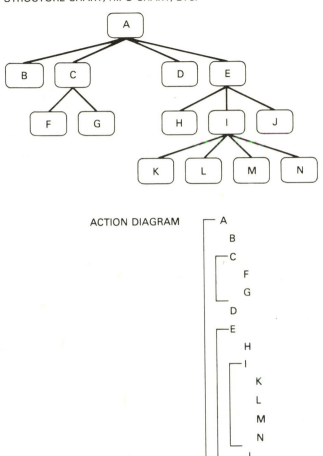

Figure 16.1 Hierarchical block structure and the equivalent action diagram.

ULTIMATE DECOMPOSITION

Figure 16.2 illustrates an action diagram overview of a program structure. Figure 16.2 is equivalent to the Warnier–Orr chart of Fig. 11.5, the Jackson chart of Fig. 12.3, the structure chart of Fig. 9.1, or the HIPO diagram of Fig. 10.1. Unlike these charts, however, it can be extended to show conditions, case structures, and loops of different types; it can show detailed program logic.

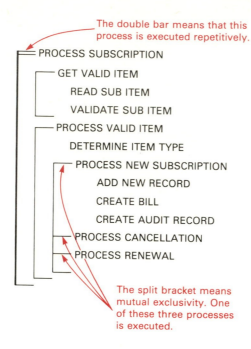

The double bar means that this process is executed repetitively.

PROCESS SUBSCRIPTION

GET VALID ITEM
READ SUB ITEM
VALIDATE SUB ITEM

PROCESS VALID ITEM
DETERMINE ITEM TYPE
PROCESS NEW SUBSCRIPTION
ADD NEW RECORD
CREATE BILL
CREATE AUDIT RECORD
PROCESS CANCELLATION
PROCESS RENEWAL

The split bracket means mutual exclusivity. One of these three processes is executed.

Figure 16.2 High-level action diagram, equivalent to the structure chart in Fig. 9.1, the HIPO chart in Fig. 10.1, the Jackson diagram in Fig. 12.3, or the Warnier–Orr chart in Fig. 11.5. This action diagram can now be expanded into a chart showing the detailed program logic. VALIDATE SUB ITEM from this chart is expanded into detailed logic in Fig. 16.3.

Figure 16.3 expands the process in Fig. 16.2 called VALIDATE SUB ITEM. Figures 16.2 and 16.3 could be merged into one chart. Figure 16.3 is equivalent to the flowchart of Fig. 13.2, the structured English of Fig. 14.1, or the Nassi–Shneiderman chart of Fig. 15.1. Glancing ahead, Fig. 16.7 shows *executable* program code written in a fourth-generation language.

This diagramming technique can thus be extended all the way from the highest-level overview to working code in a fourth-generation language. When it is used on a computer screen, the developers can edit and adjust the diagram and successively fill in detail until they have working code that can be tested interpretively. We refer to this as *ultimate decomposition*. As we will see later, the process of ultimate decomposition can be linked to data-base planning and design.

CONDITIONS

Often a program module or subroutine is executed only IF a certain condition applies. In this case the condition is written at the head of a bracket, thus:

IF CUSTOMER# IS VALID

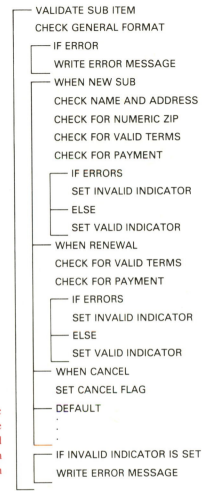

Figure 16.3   Action diagram showing the detailed logic inside the process VALIDATE SUB ITEM. This is the logic shown in the flowchart of Fig. 13.2, the structured English of Fig. 14.1, and the Nassi–Shneiderman diagram of Fig. 15.1. This diagram, showing detailed logic, is an extension of the overview diagram of Fig. 16.2.

Conditions are often used to control mutually exclusive choices:

```
      ┌─────── IF CUSTOMER# IS VALID
 ──────────
 ──────────
 ──────────
      ├─────── ELSE
 ──────────
      └
```

This has only two mutually exclusive conditions, a true and a false. Sometimes there are many mutually exclusive conditions, as follows:

```
┌─────────  WHEN KEY = "1"
├ ----------
│ ----------
├─────────  WHEN KEY = "2"
│ ----------
│ ----------
├─────────  WHEN KEY = "3"
│ ----------
│ ----------
├─────────  WHEN KEY = "4"
│ ----------
│ ----------
└
```

LOOPS

A loop is represented with a repetition bracket with the double line at its top. When many people first start to program, they make mistakes with the point at which they test a loop. Sometimes the test should be made *before* the actions of the loop are performed and sometimes the test should be made *after*. This difference can be made clear on brackets by drawing the test either at the top or bottom of the bracket, thus:

```
╒═════════  REPEAT WHILE C1        ╒═════════
│ ----------                       │ ----------
│ ----------                       │ ----------
│ ----------                       │ ----------
│ ----------                       │ ----------
└                                  └─────────  REPEAT UNTIL C1
```

If the test is at the head of the loop as with a WHILE loop, the actions in the loop may never be executed if the WHILE condition is not satisfied. If the test is at the bottom of the loop, as with an UNTIL loop, the actions in the loop are executed at least once. They will be executed more than once if the condition is fulfilled.

(Note: Some fourth generation languages put the UNTIL at the top.)

SETS OF DATA

Sometimes a procedure needs to be executed on all of the items in a set of items. It might be applied to all transactions or all records in a file; for example,

```
╒═════════  FOR ALL TRANSACTIONS
│ ----------
│ ----------
│ ----------
│ ----------
└
```

Action diagrams have been used with fourth-generation languages such as NOMAD, MANTIS, FOCUS, RAMIS, IDEAL, and so on. They are a good tool for teaching end users to work with these languages. Some of these languages have a FOR construct with a WHERE clause to qualify the FOR; for example,

FOR EACH TRANSACTION WHERE CUSTOMER# > 5000

SUBPROCEDURES

Sometimes a user needs to add an item to an action diagram which is itself a procedure that may contain actions. We call this a subprocedure, or subroutine, and draw it with a round-cornered box. A subprocedure might be used in several procedures. It will be exploded into detail, showing the actions it contains, in another chart.

SUBPROCEDURES NOT YET DESIGNED

In some cases the procedure designer has sections of a procedure which are not yet thought out in detail. He can represent this as a box with rounded corners and a wavy right edge:

COMMON PROCEDURES

Some procedures appear more than once in an action diagram because they are called (or invoked) from more than one place in the logic. These procedures are called *common procedures*. They are indicated by drawing a vertical line down the left-hand side of the procedure box, as follows:

The use of procedure boxes enables an action diagrammer to concentrate on those parts of a procedure with which he is familiar. Another person may, perhaps, fill in the details in the boxes. This enables an elusive or complex procedure formation problem to be worked out a stage at a time.

The use of these boxes makes action diagrams a powerful tool for designing procedures at many levels of abstraction. As with other structured techniques, top-down design can be done by first creating a gross structure with such boxes, while remaining vague about the contents of each box. The gross structure can then be broken down into successive levels of detail. Each explosion of a box adds another degree of detail, which might itself contain actions and boxes.

Similarly, bottom-up design can be done by specifying small procedures as action diagrams whose names appear as boxes in higher-level action diagrams.

TERMINATIONS

Certain conditions may cause a procedure to be terminated. They may cause the termination of the bracket in which the condition occurs or they may cause the termination of multiple brackets. Terminations are drawn with an arrow to the left through one or more brackets, as follows:

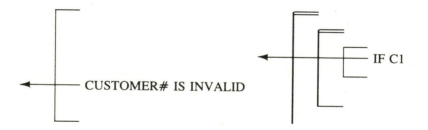

FOURTH-GENERATION LANGUAGES

When fourth-generation languages are used, the wording on the action diagram may be the wording that is used in coding programs with the language. Examples of this are as follows:

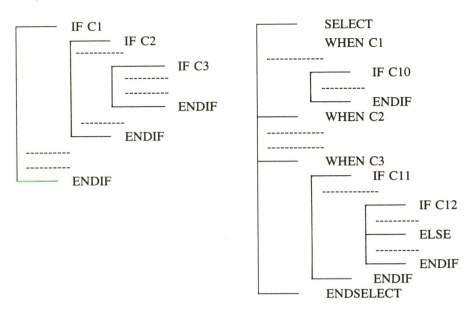

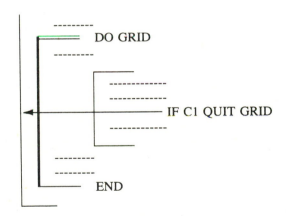

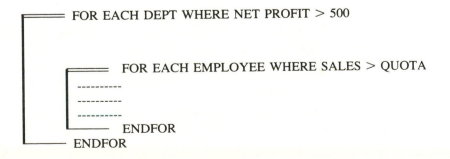

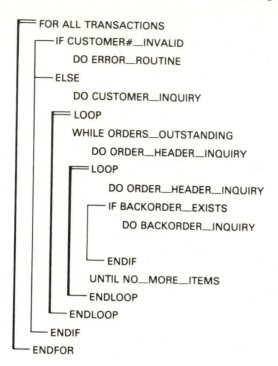

```
FOR ALL TRANSACTIONS
   IF CUSTOMER#__INVALID
         DO ERROR__ROUTINE
   ELSE
         DO CUSTOMER__INQUIRY
      LOOP
         WHILE ORDERS__OUTSTANDING
            DO ORDER__HEADER__INQUIRY
         LOOP
               DO ORDER__HEADER__INQUIRY
            IF BACKORDER__EXISTS
               DO BACKORDER__INQUIRY

            ENDIF
         UNTIL NO__MORE__ITEMS
         ENDLOOP
      ENDLOOP
   ENDIF
ENDFOR
```

Figure 16.4   Action diagrams can be labeled with the control statement of fourth-generation language and form an excellent way to teach such languages. This example uses statements from the language IDEAL from ADR [1].

Figure 16.4 shows an action diagram for a procedure using control statements from the language IDEAL for ADR [1].

DECOMPOSITION TO PROGRAM CODE

Figure 3.21 shows a Jackson diagram of a game program. With action diagrams we can decompose this until we have program code. Figure 16.5 shows an action diagram equivalent to Fig. 3.21. The action diagram gives more room for explanation. Instead of saying PRINT RANDOM WORDS it says PRINT RANDOM WORD FROM EACH OF THE THREE LISTS.

Figure 16.6 decomposes the part of the diagram labeled BUZZWORD GENERATOR. The inner bracket is a repetition bracket that executes 22 times. This is inside a bracket which is terminated by the operator pressing the ESC (escape) key. The last statement in this bracket is WAIT, indicating that the system will wait after executing the remainder of the bracket until the operator presses the ESC key. This gives the operator as much time as he wants to read the printout.

Figure 16.7 decomposes the diagram further into an executable program. This program is written in the fourth-generation language MANTIS, from Cincom Systems, Inc. [2].

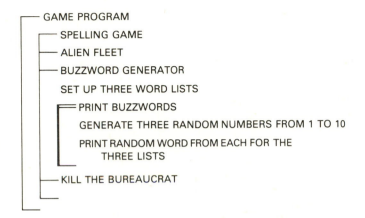

Figure 16.5   Action diagram equivalent to the Jackson diagram of Fig. 3.21.

INPUT AND OUTPUT DATA   The brackets of the action diagram are quick and easy to draw. If the user wants to show the data that enter and leave a process, the bracket is expanded into a rectangle, as shown in Fig. 16.10. The data entering the process are written at the top right-hand corner of the block. The data leaving are written at the bottom right-hand corner.

Rectangles can be nested as they were in Fig. 16.10. We then have species II functional decomposition, as described in Chapter 8. This type of functional decomposition is designed for computerized checking to ensure that all of the inputs and outputs balance.

```
┌─ BUZZWORD GENERATOR
│   PRINT INSTRUCTION TO OPERATOR
│   SET UP ADJECTIVE1 LIST
│   SET UP ADJECTIVE2 LIST
│   SET UP NOUN LIST
│  ┌═ DO
│  │   COUNT = 1
│  │  ┌═ WHILE COUNT < 22
│  │  │   GENERATE 3 RANDOM NUMBERS FROM 1 TO 10
│  │  │   PRINT  ADJECTIVE1,  ADJECTIVE2,  NOUN
│  │  └   ADD 1 TO COUNT
│  │   WAIT
└  └─  UNTIL OPERATOR PRESSES "ESC"
```

Figure 16.6   Expansion of the buzzword-generator portion of Fig. 16.5.

```
┌─ ENTER BUZZWORD GENERATOR
│
│     CLEAR
│     SHOW "I WILL GENERATE A SCREEN FULL OF 'BUZZ PHRASES' EVERY
│     TIME YOU HIT 'ENTER'. WHEN YOU WANT TO STOP, HIT 'ESC'."
│
│     TEXT ADJECTIVE1 (10,16), ADJECTIVE (10,16) NOUN (10,16)
│
│     ADJECTIVE1 (1) = "INTEGRATED," "TOTAL," "SYSTEMATIZED," PARALLEL,"
│       "'FUNCTIONAL," "RESPONSIVE," "OPTIONAL," "SYNCHRONIZED,"
│       "'COMPATIBLE," BALANCED"
│
│     ADJECTIVE2 (1) = "MANAGEMENT," "ORGANIZATIONAL," "MONITORED,"
│       "'RECIPROCAL," "DIGITAL," "LOGISTICAL," "TRANSITIONAL,"
│       "'INCREMENTAL," "THIRD GENERATION," "POLICY"
│
│     NOUN(1) = "OPTION," "FLEXIBILITY," "CAPABILITY," "MOBILITY,"
│       "'PROGRAMMING," "CONCEPT," "TIME PHASE," "PROJECTION,"
│       "'HARDWARE," "CONTINGENCY"
│
│     SEED
│  ┌═ WHILE KEY ≠ "ESC"
│  │  COUNT = 1
│  │
│  │  ┌═ WHILE COUNT < 22
│  │  │  A = INT(RND(10) + 1)
│  │  │  B = INT(RND(10) + 1)
│  │  │  C = INT(RND(10) + 1)
│  │  │
│  │  │  SHOW ADJECTIVE1(A) + " " + ADJECTIVE2(B) + " " + NOUN(C)
│  │  │
│  │  │  COUNT = COUNT + 1
│  │  └─ END
│  │
│  │     WAIT
│  └─ END
│
│     CHAIN "GAMES_MENU"
└─ EXIT
```

Figure 16.7　Expansion of the action diagram of Fig. 16.6 into program code. This is an executable program in the fourth-generation language MANTIS [2]. Successive decomposition of a diagram until it becomes executable code is called *ultimate decomposition*.

The square brackets may be thought of as a shorthand way of drawing rectangles like those in Fig. 16.10.

TITLES VERSUS CODE STRUCTURE

At the higher levels of the design process, action diagram brackets represent the *names* of processes and subprocesses. As the designer descends into program-level detail, the brackets become *program constructs*: IF brackets, CASE brackets,

LOOP brackets, and so on. To show the difference, different colors may be used. The *name* brackets may be *red* and the *program construct black*. If a black-and-white copier or terminal is used, the *name* brackets may be dotted or gray and the *program-construct* brackets black.

The program-construct brackets may be labeled with appropriate control words. These may be the control words of a particular programming language, or they may be language-independent words.

Figure 16.8 shows the program constructs with language-independent control words. Figure 16.9 shows the same constructs with the words of the fourth-generation language IDEAL [1]. It is desirable that any fourth-generation language should have a set of clear words equivalent to Fig. 16.1, and it would help if standard words for this existed in the computer industry.

Simple action diagram editors exist for personal computers [3], which speed up the production and modification of programs and help eliminate common types of bugs. A panel for the rapid adding or modification of brackets is shown in Fig. 16.10. A computer can select the control words for any chosen language. In the examples in this book the control words are shown in boldface type. The other statements are shown in regular type.

A diagramming technique today should be designed for both quick manual manipulation and for computerized manipulation. Users and analysts will want to draw rough sketches on paper or argue at a blackboard using the technique. They will also want to build complex diagrams at a computer screen, using the computer to validate, edit, and maintain the diagrams, possibly linking them to a dictionary, data-base model, and so on. The tool acts rather like a word processor for diagramming, making it easy for users to modify their diagram. Unlike a word processor, it can perform complex validation and cross-checking on the diagram.

The design of simple programs does not need automated correlation in inputs and outputs, or diagrams like Fig. 16.11, which show the inputs and outputs. In the design of complex specifications the automated correlation of inputs and outputs among program modules is essential if mistakes are to be avoided.

In showing input and output data, Fig. 16.11 contains the information on a data flow diagram. It can be converted into a layered data flow diagram as in Fig. 16.12. Unlike a data flow diagram, it can be directly extended to show the program structure, including conditions, case constructs, and loop control.

We think it highly desirable that a programmer sketch the structure of his programs with action diagram brackets. These can be drawn on the coding sheet. The author has checked many programs written with fourth-generation languages such as FOCUS, RAMIS, IDEAL, NATURAL, and so on. Often the coder has made a logic error in his use of loops, END statements, CASE structures, EXITS, and so on. When he is taught to draw action diagram brackets and fit the code to them, these structure errors become apparent. The control statements can be fitted to the brackets. Figure 16.13 shows a program written in the C language designed with an action diagram.

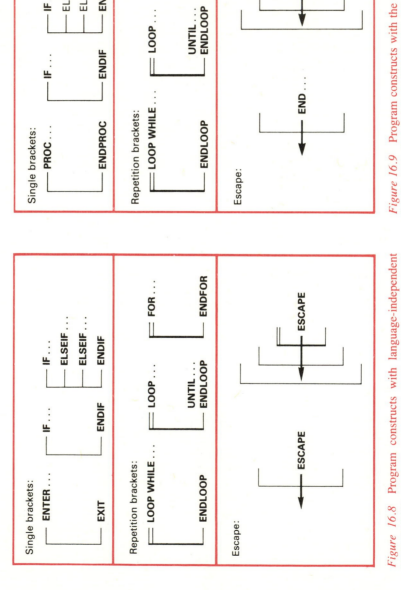

Figure 16.8 Program constructs with language-independent control words.

Figure 16.9 Program constructs with the control words of the fourth-generation language IDEAL.

Single brackets:

     Block:☐   Condition:☐

     Case: 1:☐   2:☐   3:☐   4:☐   More:☐

Repetition brackets:

     For all:☐   Loop while:☐   Loop until:☐

Escape:

     1:☐   2:☐   3:☐   4:☐   5:☐   More:☐

Figure 16.10   Menu for rapid addition or modification of brackets on an action diagram.

Software, which can run like word-processing software on personal computers, can be used for building, editing, and modifying action diagrams, and fitting code to them.

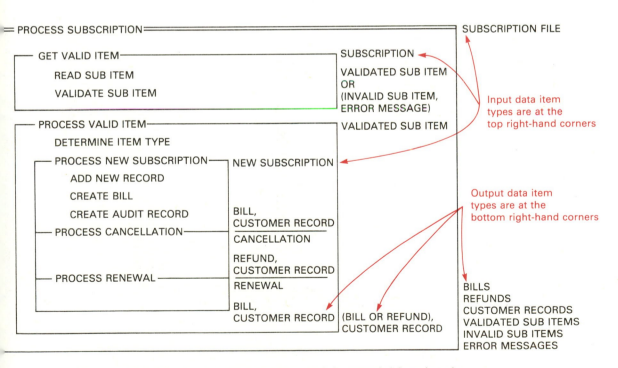

Figure 16.11   The bracket format of Fig. 16.2 is expanded here into the rectangular format used to show the data-item types which are input and output to each process. This is designed for computerized cross-checking.

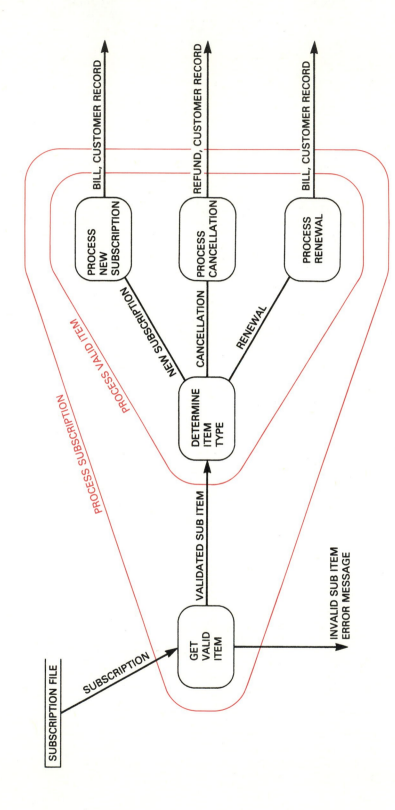

Figure 16.12 Data flow diagram corresponding to Fig. 16.11.

198

```
+- addel ()
|
|       int itab;
|       int adlevel;
|       int itfrst, itnext, itdel;
|
|     +- switch (adtab[curline].action) {
|     |
|     +- case ACT_JUNK:
|     +- case ACT_CASW:
|     +- case ACT_EXIT:
|     |       itfrst = curline;
|     |       itnext = curline + 1;
|     |       break;
|     |
|     +- case ACT_BEGL:
|     +- case ACT_BEGB:
|     |       adlevel = 1;
|     |       itfrst = curline;
|     |       for (itnext = curline+1; itnext <= numadtab & adlevel > 0; itnext++)
|     |       +- switch (adtab[itnext].action) {
|     |       +- case ACT_BEGL:
|     |       +- case ACT_BEGB:
|     |       |       adlevel++;
|     |       |       break;
|     |       +- case ACT_END:
|     |       |       adlevel--;
|     |       +-     }
|     |           }
|     |     +- if (adlevel > 0)
|     |     |       aborts ("ad: ? no matching end ??");
|     |     +-
|     |       break;
|     |
|     +- default:
|     |       itfrst = numadtab;
|     |       itnext = numadtab;
|     |       beep();
|     +-     }
|
|       itdel = itnext - itfrst;
|       for (itab = itfrst; itab < numadtab; itab++) {
|           adtab[itab].action = adtab[itab+itdel].action;
|           adtab[itab].count  = adtab[itab+itdel].count;
|           adtab[itab].text   = adtab[itab+itdel].text;
|           }
|       numadtab -= itdel;
|     +- if (curline > numadtab)
|     |       curline = numadtab;
|     +-
|       showbuffer ();
+-
```

Figure 16.13 An action diagram used with the language C. The code was edited with the action diagram editor on the screen of an IBM personal computer. This is a printout from the PC.

AUTOMATIC DERIVATION OF ACTION DIAGRAMS

Action diagrams can be derived automatically from correctly drawn decomposition diagrams (Chapter 5), dependency diagrams (Chapter 6), data navigation charts (Chapter 21) or decision trees (Chapter 17). If a computer algorithm is used for doing this, it needs to check the completeness or integrity of the dependency diagram or navigation chart. This helps to validate or improve the analyst's design.

Figure 16.14 gives examples of decomposition diagrams and their corresponding action diagrams. Figure 16.15 gives examples of dependency diagrams and their corresponding action diagram. Later in the book, Fig. 21.2 gives examples of data navigation charts and their corresponding action diagram.

It should be noted that many types of drawings which analysts create *cannot* be converted *automatically* in action diagrams. When this is the case, it represents a serious defect in the methodology. It is usually desirable to abandon methodologies that do not permit automatic conversion to action diagrams or their equivalent.

CONVERSION OF ACTION DIAGRAMS TO CODE

Different computer languages have different commands relating to the constructs drawn on action diagrams. If the action diagram is being edited on a computer screen, a menu of commands can be provided for any particular language. Using the language IDEAL, for example, the designer might select the word LOOP for the top of a repetition bracket, and the software automatically puts END-LOOP at the bottom of the bracket and asks the designer for the loop control statement. The designer might select IF and the software creates the following bracket.

The user is asked to specify the IF condition.

Such structures with the commands for a given language may be automatically generated from a dependency diagram or data navigation chart. The objective is to speed up as much as possible the task of creating error-free code.

With different menus of commands for different languages, a designer may switch from one language to another, if necessary. This facilitates the adoption of different languages in the future.

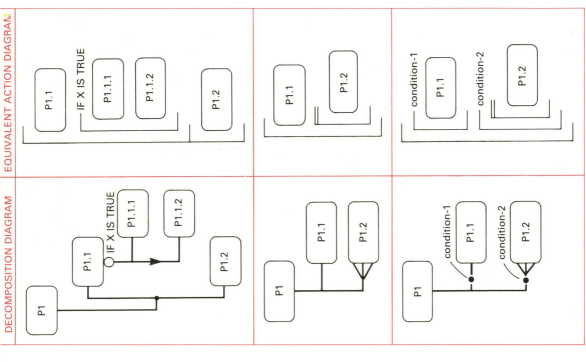

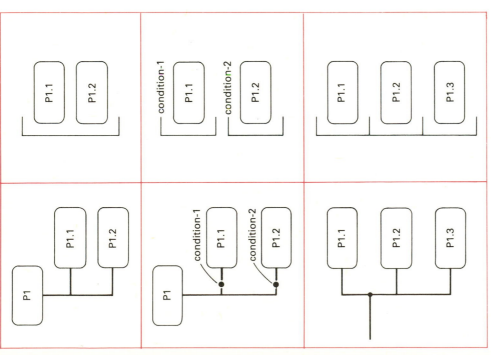

Figure 16.14 Decomposition diagrams and their equivalent action diagram.

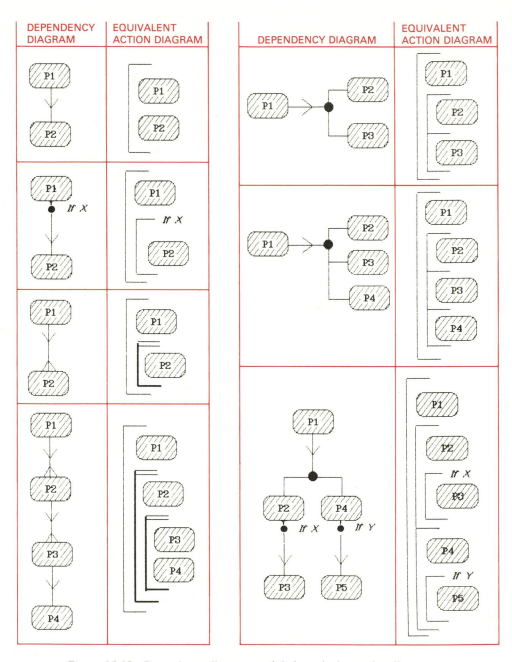

Figure 16.15 Dependency diagrams and their equivalent action diagram.

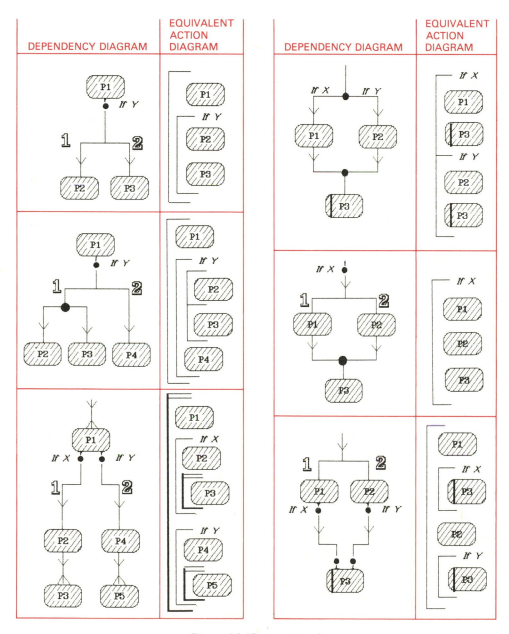

Figure 16.15   (continued)

BOX 16.1 Summary of notation used in action diagrams

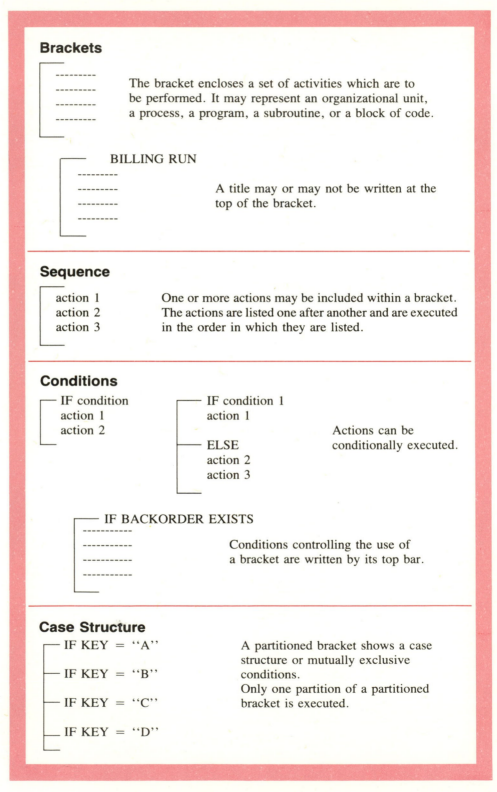

Brackets

The bracket encloses a set of activities which are to be performed. It may represent an organizational unit, a process, a program, a subroutine, or a block of code.

BILLING RUN

A title may or may not be written at the top of the bracket.

Sequence

action 1
action 2
action 3

One or more actions may be included within a bracket. The actions are listed one after another and are executed in the order in which they are listed.

Conditions

IF condition
action 1
action 2

IF condition 1
action 1

ELSE
action 2
action 3

Actions can be conditionally executed.

IF BACKORDER EXISTS

Conditions controlling the use of a bracket are written by its top bar.

Case Structure

IF KEY = "A"

IF KEY = "B"

IF KEY = "C"

IF KEY = "D"

A partitioned bracket shows a case structure or mutually exclusive conditions. Only one partition of a partitioned bracket is executed.

BOX 16.1 · *(Continued)*

Repetition

A double bar at the top of a bracket indicates that the contents of the bracket will be executed multiple times; for example, it is used to draw a program loop.

DO WHILE N > O   Conditions controlling a DO WHILE loop are written at the top of the bracket, showing that the condition is tested before the contents of the bracket are executed.

DO   Conditions controlling a DO UNTIL loop are written at the bottom of the bracket, showing that the condition is tested after the contents of the bracket are executed.

UNTIL NO MORE ITEMS

FOR ALL . . .

FOR EACH . . .
WHERE . . .

Nesting

Brackets are nested to show a hierarchy— a form of tree structure.

BOX 16.1 *(Continued)*

Rectangle Format

The bracket may be expanded into a rectangle. The inputs to the activities in the rectangle are written at its top left-hand corner; the outputs are written at its bottom right-hand corner.

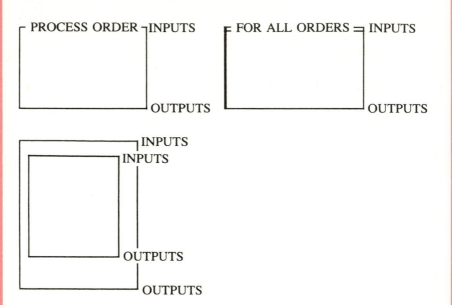

The rectangle format is designed for use with a computer which assists in drawing and cross-checks the inputs and outputs.

Exits

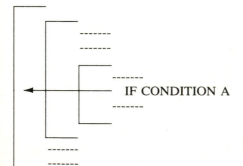

IF CONDITION A

An arrow to the left through one or more brackets indicates that the brackets it passes through are terminated if the condition written by the arrow is satisfied.

BOX 16.1 *(Continued)*

Subprocedures

A round-cornered box within a bracket indicates a procedure diagrammed elsewhere.

A round-cornered box with a broken edge indicates a procedure not yet thought out in more detail.

Common Procedures

A procedure box with a vertical line drawn through the left side indicates a common procedure— that is, a procedure that appears multiple times in the action diagram.

The following relate to data-base action diagrams which are discussed in Chapters 21 and 22.

Simple Data Action

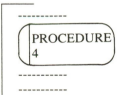

A rectangle containing the name of a record type or entity type is preceded by a simple data access action: CREATE, READ, UPDATE, or DELETE.

BOX 16.1 *(Continued)*

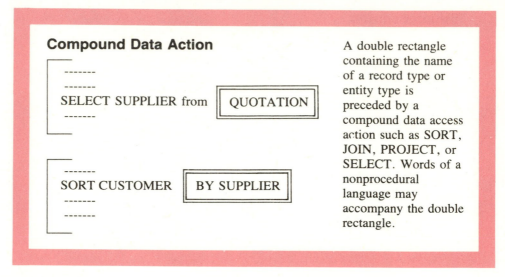

Compound Data Action

SELECT SUPPLIER from   QUOTATION

SORT CUSTOMER   BY SUPPLIER

A double rectangle containing the name of a record type or entity type is preceded by a compound data access action such as SORT, JOIN, PROJECT, or SELECT. Words of a nonprocedural language may accompany the double rectangle.

ADVANTAGES   Action diagrams were designed to solve some of the concerns with other diagramming techniques. They were designed to have the following properties:

1. They are quick and easy to draw and to change.

2. They are good for manual sketching and for computerized editing.

3. A single technique extends from the highest overview down to coding-level detail (ultimate decomposition).

4. They draw all the constructs of traditional structured programming and are more graphic than pseudocode.

5. They are easy to teach to end users; they encourage end users to extend their capability into examination or design of detailed process logic. They are thus designed as an information center tool.

6. They can be printed on normal-width paper rather than wall charts, making them appropriate for design with personal computers.

7. Data navigation diagrams (Chapter 21) can be converted *automatically* into action diagrams. These can include compound relational operations (Chapter 22).

8. Action diagrams are designed to link to a data model.

9. They work well with fourth-generation languages and can be tailored to a specific language dialect.

10. They are designed for computerized cross-checking of data usage on complex *specifications*.

Some aspects of the list above did not exist when the early structured techniques were designed:

- Computerized editing of an analyst's diagrams
- Computerized cross-checking
- Data models
- Compound relational data-base operations
- Use of personal computers for design
- Fourth-generation languages
- Strong end-user involvement in computering
- Information center management

Chapter 21 extends action diagramming techniques to show data-base operations. Box 16.1 summarizes the diagramming conventions of action diagrams.

REFERENCES

1. IDEAL manual from Applied Data Research Inc., Princeton, N.J.

2. The example in Figs. 16.6 and 16.7 is adapted from a program in the *MANTIS User's Guide*, Cincom Systems, Inc., Cincinnati, Ohio, 1982.

3. Action Diagrammer, from DDI, Ann Arbor, MI.

17 DECISION TREES AND DECISION TABLES

A BROADLY USED DIAGRAMMING TECHNIQUE

Decision trees and decision tables did not originate as computer diagramming techniques. They have much broader applicability. Decision trees and decision tables are used in biology, computer science, information theory, and switching theory. There are four main fields of application [1]:

1. Taxonomy, diagnosis, and pattern recognition

2. Circuit (logic) design and reliability testing

3. Analysis of algorithms

4. Decision table programming and data bases

DECISION TREE

A *decision tree* is a model of a discrete function in which the value of a variable is determined, and then based on this value some action is taken [1]. The action is either to choose another variable to evaluate or to output the value of the function. Thus each action taken depends on the current value of the variable being tested and all previous actions that have been taken. In a formally defined decision tree, a variable is tested only once on any path through the tree. This restriction is to prevent redundant testing.

Decision trees are normally constructed from a problem description. They give a graphic view of the decision making that is needed. They specify what variables are tested, what actions are to be taken, and the order in which decision making is performed. Each time a decision tree is "executed" one path, beginning with the root of the tree and ending with a leaf on the tree, will be followed depending on the current value of the variable(s) tested.

Consider the description for the subscription system:

In the subscription system, subscription transactions are processed. First, each transaction is validated. Invalid transactions are rejected with an appropriate error message. Valid transactions are processed depending on their type: new subscription, renewal, or cancellation. For new subscriptions, a customer record is built and a bill is generated for the balance due. For renewals, the expiration data are updated and a bill is generated for the balance due. For cancellations, the record is flagged for deletion and a refund is issued.

A decision tree for the subscription system is shown in Fig. 17.1. The root of the tree is the condition test VALID TRANSACTION?, which is answered "yes" or "no." If a transaction is valid and it is a new subscription, the path with a "yes" to the VALID TRANSACTION? condition test and a "new subscription" for TRANSACTION TYPE will be followed. This path ends with processing a new subscription by building a customer record and generating a bill.

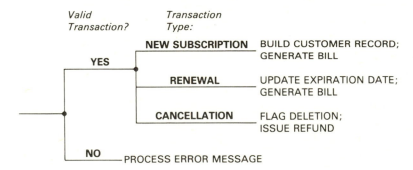

Figure 17.1   A decision tree is a graphic view of the decision logic in a program function.

Figure 17.2 shows a more complex (and more useful) decision tree. It relates to complex rules for determining how much discount a customer receives. At the top of the diagram are the names of data attributes. CUSTOMER.ANNUAL PURCHASES, for example, refers to the data item ANNUAL PURCHASES in the CUSTOMER record. If this has a value greater than 1 million, the bottom part of the tree is used. The values of four attributes are used to determine ORDER.DISCOUNT.

Mutual Exclusivity

The paths on a decision tree are normally mutually exclusive. The dot on the branch, which we employed earlier in the book to represent mutual exclusivity, is shown on our drawings of decision trees.

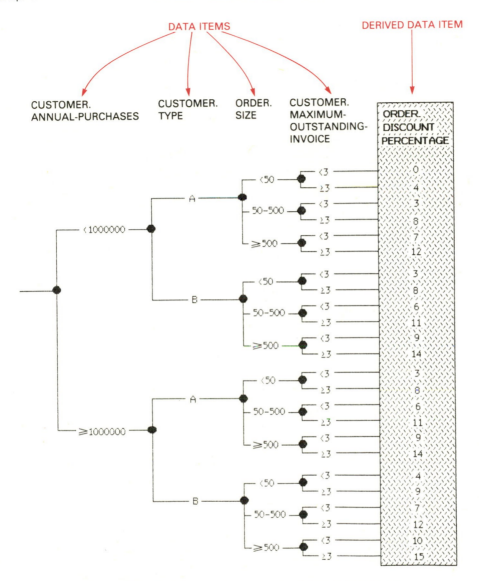

Figure 17.2 Decision tree for computing ORDER DISCOUNT.

Translation into Action Diagrams

Decision trees showing a choice of processes translate directly into action diagrams. Figure 17.3 shows an action diagram which is equivalent to Fig. 17.1. The shape of a *case* structure bracket is similar to the shape of a branching line in the decision tree.

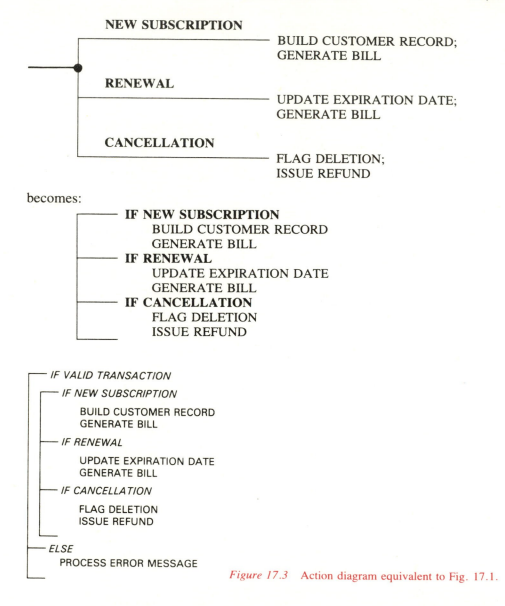

becomes:

Figure 17.3   Action diagram equivalent to Fig. 17.1.

DECISION TABLE   A *decision table* is an alternative model for a function. It shows the function in a tabular or matrix form where the upper rows of the table specify the *variables* or *conditions* to be evaluated and the lower rows specify the *corresponding action* to be taken when an evaluation test is satisfied. A column in the table is called a *rule*. Each rule defines a procedure of the type: If the condition is true, then execute the corresponding action.

CONDITIONS				
Valid transaction	NO	YES	YES	YES
New subscription	—	YES	NO	NO
Renewal	—	NO	YES	NO
Cancellation	—	NO	NO	YES
ACTIONS				
Process error message	x			
Build customer record		x		
Generate bill		x	x	
Update expiration date			x	
Flag deletion				x
Issue refund				x

Figure 17.4 A decision table can give a tabular view of the decision logic in a program. This table is equivalent to the decision tree in Fig. 17.1.

A column is referred to as a "rule"

Figure 17.4 shows a decision table for the subscription system. The rule for processing a new subscription (the second column) tells us that if the transaction is valid, the actions to take are to build a customer record and generate a bill.

A dash in the condition row means "do not care." In the first column of Fig. 17.4 the dashes mean that nothing else matters if the transaction is invalid.

The decision table in Fig. 17.4 relates to relatively simple decisions. They are sufficiently simple that other forms of diagram are probably better and relate more directly to program structure; for example, the action diagram shown in Fig. 17.3.

However, if the decisions relate to complex combinations of conditions, a decision table is a major aid to clear thinking. Figure 17.5 shows a decision table where this is the case. Some software exists that can convert decision tables such as Fig. 17.5 into program code.

An attempt to draw the logic of Fig. 17.5 with flowcharts, Nassi–Shneiderman charts, Warnier–Orr charts, or similar methods would be clumsy, take a long time, and the designer would probably make errors.

DECISION TREE OR TABLE?
When should you use decision trees and when decision tables? Decision trees are easier to read and understand when the number of conditions is small. Most persons would grasp the meaning of Fig. 17.1 without special training, for example. They might be more bewildered by Fig. 17.4.

If there are a substantial number of conditions and actions, a decision tree becomes too big and clumsy. A decision tree for the situation in Fig. 17.5 would be too large. The patterns in a table like Fig. 17.5 give a clearer idea what is required and encourage visual checking.

CONDITIONS:

			P1						P2						P3						P4						P5						P6						P7				
STATE OF THE INTERFACE																																											
TYPE OF MESSAGE RECEIVED	CALL REQUEST	CALL ACCEPTED	CLEAR REQUEST	CLEAR CONFIRM	DATA	RESET	CALL REQUEST	CALL ACCEPTED	CLEAR REQUEST	CLEAR CONFIRM	DATA	RESET	CALL REQUEST	CALL ACCEPTED	CLEAR REQUEST	CLEAR CONFIRM	DATA	RESET	CALL REQUEST	CALL ACCEPTED	CLEAR REQUEST	CLEAR CONFIRM	DATA	RESET	CALL REQUEST	CALL ACCEPTED	CLEAR REQUEST	CLEAR CONFIRM	DATA	RESET	CALL REQUEST	CALL ACCEPTED	CLEAR REQUEST	CLEAR CONFIRM	DATA	RESET	CALL REQUEST	CALL ACCEPTED	CLEAR REQUEST	CLEAR CONFIRM	DATA	RESET

ACTIONS:

Action \ State	P1 CR	P1 CA	P1 ClR	P1 CC	P1 D	P1 RS	P2 CR	P2 CA	P2 ClR	P2 CC	P2 D	P2 RS	P3 CR	P3 CA	P3 ClR	P3 CC	P3 D	P3 RS	P4 CR	P4 CA	P4 ClR	P4 CC	P4 D	P4 RS	P5 CR	P5 CA	P5 ClR	P5 CC	P5 D	P5 RS	P6 CR	P6 CA	P6 ClR	P6 CC	P6 D	P6 RS	P7 CR	P7 CA	P7 ClR	P7 CC	P7 D	P7 RS
PROCEDURE 1	X																																									
PROCEDURE 2								X																																		
PROCEDURE 3									X																																	
PROCEDURE 4														X																												
PROCEDURE 5													X																													
ERROR MESSAGE A		X																																								
ERROR MESSAGE B			X																							X						X						X				
ERROR MESSAGE C				X											X						X						X						X						X			
ERROR MESSAGE D					X											X						X						X						X						X	
ERROR MESSAGE E						X						X						X	X	X																						
PRINT INSTUCTION K																																					X					
ERROR PROCEDURE 1	X	X	X	X	X	X				X	X	X			X	X		X			X	X	X	X	X	X		X		X	X	X		X		X		X				X
ERROR PROCEDURE 2																			X	X	X	X	X	X				X		X	X	X				X	X	X				X
LOG OPERATION					X						X						X						X												X							
SECURITY PROCEDURE																																										
SEE DECISION TABLE 13																																										
SEE DECISION TABLE 14																																										
SEE DECISION TABLE 15																													X	X					X	X					X	
SEE DECISION TABLE 16																																									X	X

Figure 17.5 Whereas simple decisions such as those in Fig. 17.4 can be handled by other methods, complex decisions such as those illustrated here require a decision table. Program code can be generated automatically from decision tables.

216

A decision table like Fig. 17.5 causes the designer to look at every possible combination. Without such a technique he would probably miss certain combinations. The decision tree does not provide a matrix for every condition and action. It is thus easier to omit important combinations when using a decision tree.

USING DECISION TREES AND DECISION TABLES

Because a decision table (tree) maps inputs (conditions) to outputs (actions) without necessarily specifying how the mapping is to be done, it has been used as a system analysis and system design tool. In particular, it has been used to describe the decision-making logic in a functional specification and the program control structure in a program design. For example, DeMarco recommends that in structured systems analysis, the process specification can be described by a decision table (tree). This is an alternative to using pseudocode, which is not as easy to understand.

Traditionally, flowcharts have been used to represent detailed and complex logic graphically. However, decision tables or trees are preferable because they offer a more compact view of the logic.

Not only can decision tables (trees) model program control logic, but they can also model an entire program. Theoretically, they can represent any computable function and therefore replace any program flowchart [2]. The control constructs of sequence, selection, and repetition can be represented by a decision table or tree.

COMMENTARY

Decision tables and decision trees are used as detail design tools for complex program logic. They are seldom used at a high level to show program control structure. In general, they should not be used as a stand-alone design tool. Instead, they should be used to supplement other design tools. They can be converted *automatically* into action diagrams.

Decision tables are a valuable tool, especially when used with software which automatically converts them to program code. They have been surprisingly neglected by many DP organizations.

REFERENCES

1. Moret, B., "Decision Trees and Diagrams," *Computing Surveys*, Vol. 14, No. 4 (December 1982): 593–623.

2. Lew, A., "In the Emulation of Flowcharts by Decision Tables," *CACM*, Vol. 25, No. 12 (December 1982): pp. 895–905.

18 STATE-TRANSITION DIAGRAMS

INTRODUCTION　Decision trees and tables provide a valuable means of designing certain types of logic. This chapter describes another tool for designing certain categories of logic—finite-state machine notation. It is useful where entity types, switches, or variables can be thought of as being in a given number of states or complex logic governs the transitions among these states. It has been used in the design of control program mechanisms, systems software, and computer network protocols. It has much wider applicability, but is generally not understood by many systems analysts.

FINITE-STATE MACHINES　A finite-state machine is a hypothetical mechanism which can be in one of a discrete number of conditions of states. Certain events can cause it to change its state. A process can be represented as a collection of finite-state machines. This gives a precise way to draw and conceptualize complex processes and to check that all possible state transitions have been thought about.

DISCRETE POINTS IN TIME　In this view of the world, events occur at discrete points in time. There is no slow, continuous change. Events cause an instantaneous change in the state of a finite-state machine. The events can occur asynchronously, that is, at any point in time, or synchronously, at clock intervals.

State-transition diagrams are drawn to represent the behavior of finite-state machines. In the common way of drawing these, the possible states are drawn as circles.

To illustrate a state-transition diagram, Fig. 18.1 shows a dismal view of the life of a man. The man is regarded as a finite-state machine which can be in

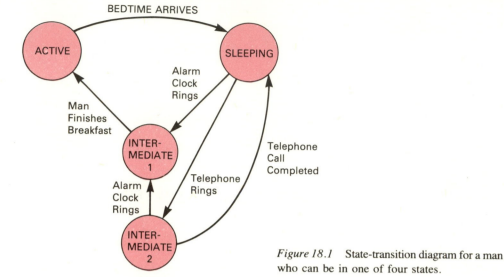

Figure 18.1 State-transition diagram for a man who can be in one of four states.

one of four states: active, sleeping, and two intermediate states. Various events cause transitions between the states, as shown by the arrows. Two intermediate states are needed because the period in those states terminates in different ways.

State-transition diagrams are commonly employed for representing those complex protocols which are needed in computer networks. Figure 18.2 shows a diagram with five states which is used to define a standard interface to packet-switching networks [1]. The diagram refers to the interface between a data processing machine or terminal and the network transport subsystem. The former is referred to as the DTE (data terminal equipment) and the latter as the DCE (data communications equipment). The diagram relates to the initiation of a call, and the consequent change of the interface state from READY to DATA TRANSFER. The call may be initiated by the DTE or it may be an incoming call, in which case the DCE initiates the process.

In complex system design we have to consider the errors, failures, and exception conditions that can occur. In Fig. 18.1 an error condition might be that the alarm clock fails. The man then appears to be permanently in the SLEEPING state. Again he might go to bed and find a beautiful female there. Such exception conditions greatly complicate the protocols.

MACHINES AND SUBMACHINES
A highly complex system, such as a packet-switching network of IBM's Systems Network Architecture (SNA), could be described as being an extremely elaborate finite-state machine with many states. This would not be a useful description because it would be too complex for human comprehension. Therefore, such

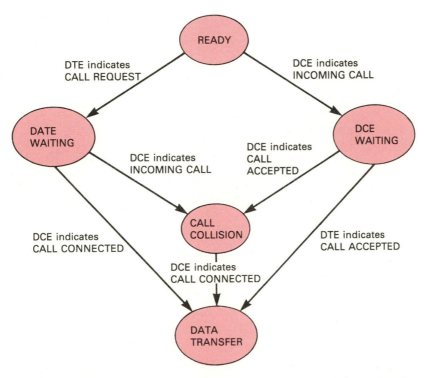

Figure 18.2　States of the DTE/DCE interface during the setup phase of a call with CCITT *Recommendation X.25* [1].

(*DTE* means *data terminal equipment*, e.g., terminal or host computer.

DCE means *data circuit-termination equipment*, e.g., network interface minicomputer). (Courtesy of IBM Corporation.)

systems are decomposed into submachines—interconnected or nested finite-state machines. The submachines can be further decomposed until we arrive at finite-state machines which are easy to comprehend, usually with not more than 10 or so states.

This is a form of functional decomposition like that described earlier. Complex processes are decomposed into subprocesses. Subprocesses are decomposed into sub-subprocesses. On a state-transition diagram a circle may be the boundary of another finite-state machine. ACTIVE, for example, in Fig. 18.1 is a highly complex state decomposable into many substates.

The state-transition diagrams used to describe the logic of distributed processing software are highly complex and appropriate fragmentation into separate modules is important. Figure 18.3 gives an illustration of how this is done for computer networking software. A session between machines is divided into layers as shown, with a finite-state machine for each layer defining the interlayer protocols. These finite-state machines are themselves subdivided into multiple, smaller machines.

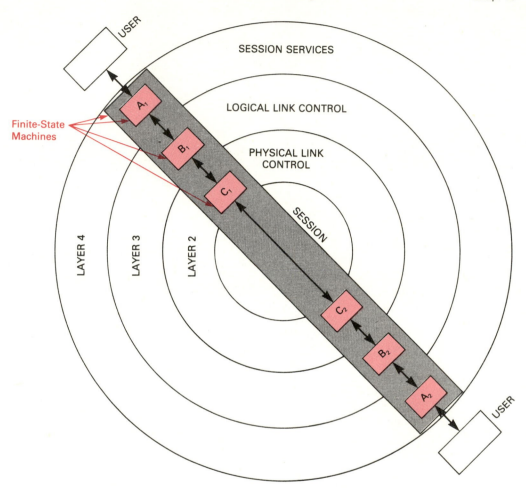

Figure 18.3 The mechanism for computer networking is divided into layers. A finite-state machine is defined for each layer as shown here. These machines define the interlayer protocols, and are themselves subdivided into multiple smaller finite-state machines.

FINITE-STATE MACHINES

A finite-state machine is thought of as a black box which can be in one of a number of states (Fig. 18.4). A finite set of input types can reach it. Inputs reach it one at a time and hence time is regarded as a set of discrete points. Two inputs cannot arrive at the same instant, although they could be infinitesimally close. It is possible that two inputs on different communication lines could arrive at *exactly* the same time, but the machine will be *scanning* its sources of input, and so will receive one before the other. Queuing mechanisms permit the inputs to be handled one at a time.

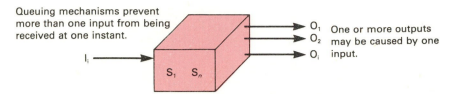

Figure 18.4 A finite-state machine has (1) a finite set of *states*, *S*; (2) a finite set of *input* types, *I*; (3) a finite set of *output* types, *O*; (4) a mapping of *input* and *present state* into *next state*; and (5) a mapping of *input* and *present state* into *output*.

In addition to causing state changes as in Figs. 18.1 and 18.2, input can also cause outputs to be sent. The input and its resulting state change and output are thought of as occurring at the same instant.

An input or output can be thought of as a *pulsed* variable which exists only at an instant in time. The state of the machine is a static discrete variable which can change only at the instant when an input is received.

Suppose that an input is received at time t_i. The state of the machine at time t_{i+1} is a function of the state $S(t_i)$ at time t_i and the input $I(t_i)$ at time t_i. We refer to this as the *next-state function*, FNS.

The output $O(t_i)$ at time t_i is also a function of the state $S(t_i)$ and input $I(t_i)$ at that time. We refer to this as the *output function*, FOUT.

$$S(t_{i+1}) = \text{FNS } (S(t_i), I(t_i))$$
$$O(t_i) = \text{FOUT } (S(t_i), I(t_i))$$

When we define these two functions we define a finite state machine.

If circles are drawn to represent the states, the arrows between them should show the inputs that cause the change of state and also the outputs that accompany it. Figure 18.5 illustrates this type of diagram. The arrows between circles, showing the state transitions, are labeled with the input that causes the transition, and with the resulting output(s).

In some cases an input causes no change of state, but does cause an output, as when input I_1 is received when the machine is in state 3. Sometimes more than one input can cause the same state transition as between states 3 and 4. Sometimes one input causes several outputs, as when state 4 received I_4.

The type of drawing shown in Fig. 18.5 is useful for human perception of protocol mechanisms. It does, however, leave a number of questions unanswered. For example, what would the machine do in state 2 if it received input I_4? This and many similar possibilities are not shown on the diagram. In order to force complete thinking about a protocol, a *state transition matrix* may be drawn instead of a diagram like Fig. 18.5.

Figure 18.6 gives a state-transition matrix which contains the same information as Fig. 18.5. The matrix shows the blanks in the protocol thinking. The

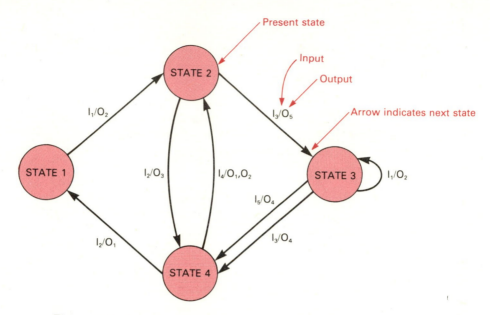

Figure 18.5   Example of a finite-state matching transition diagram, showing outputs as well as state changes.

designer should fill in all the blanks in the table. He may simply put a line through the blank spaces to indicate that certain inputs are ignored. Often, however, some form of error indication is needed if a useless input is received.

STATES DRAWN AS VERTICAL LINES

Although state-transition diagrams are normally drawn with the states represented as circles, as in Figs. 18.2 and 18.5, a neater way to draw them is to use fence diagrams, with states represented as vertical lines. Figure 18.7 shows Fig. 18.1 with the man's four states shown as vertical lines.

		STATE			
		1	2	3	4
INPUT	I_1	$2/O_2$		$3/O_2$	
	I_2		$4/O_3$		$1/O_1$
	I_3		$3/O_5$	$4/O_4$	
	I_4				$2/O_1,O_2$
	I_5			$4/O_4$	

Figure 18.6   State-transition matrix which contains the same information as that in Fig. 18.5. It is desirable, however, to fill in the blank entries.

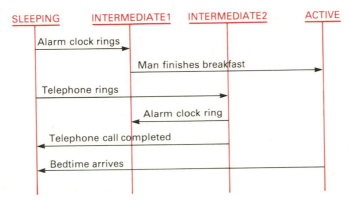

Figure 18.7 Figure 18.1 redrawn as a fence diagram. The states are shown as vertical bars. The state transitions are horizontal lines linked to the states by a dot or an arrowhead.

With complex mechanisms (like those in network architectures) there may be *many* transitions among a relatively small number of states. Bubble diagrams for showing state transitions become cluttered and confusing. Fence diagrams are much neater. Figure 18.8 shows a typical state-transition diagram in IBM's SNA. This illustrates the need for clean, formal, computer-printable state-transition diagrams.

Another reason for avoiding bubble chart state-transition diagrams is that they look like data flow diagrams, and this sometimes causes confusion about their true meaning. The circles show states and must not be confused with processes.

The horizontal lines in Figs. 18.7 and 18.8 show the input which causes the transition between states. This is written above the line. Below the line we can write the output that results when the state transition occurs. Figure 18.9 shows Fig. 18.4 redrawn in this way, with the outputs below each horizontal line. If input I_1 occurs when the system is in state 3, output O_2 occurs with no change of state. This is shown by the loop on the right.

In some cases, the input is complex; several conditions have to apply in order for a change of state to occur. In Fig. 18.8, for example, most of the horizontal lines indicate compound inputs which cause changes of state.

ENTITY LIFE CYCLES

State-transition diagrams are most commonly employed on systems which have complex state-transition logic like that found in networking software and telephone switching. They can, however, be used on commercial systems to clarify the representation of changes in state of a data entity type.

Consider the recording of a student registration by an organization that gives seminars. The registration can be in one of the following possible states: *submitted,*

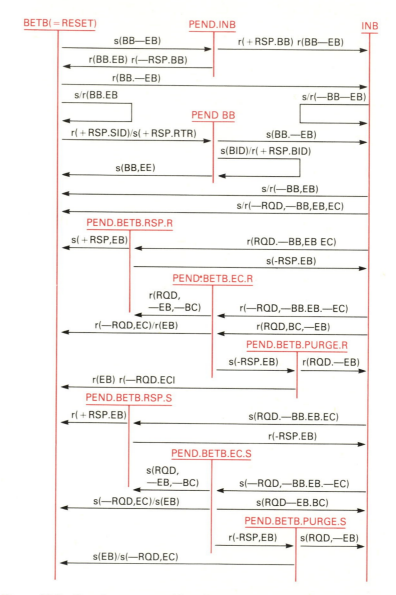

Figure 18.8 Complex state-transition diagram. This is one of many finite-state machines that define the protocols of IBM's SNA (Systems Network Architecture) [2]. Such diagrams are much clearer if drawn as fence diagrams rather than bubble charts. (Courtesy of IBM Corporation.)

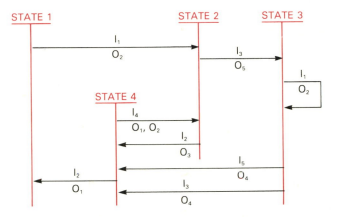

Figure 18.9   Figure 18.4 redrawn as a fence diagram.

The inputs that cause a state change are written above the horizontal lines; the outputs corresponding to the state change are written below the horizontal lines.

accepted, rejected, withdrawn, followed-up, canceled, seminar-in-progress, and *archived*. Figure 18.10 shows these states with the inputs that cause transitions between states.

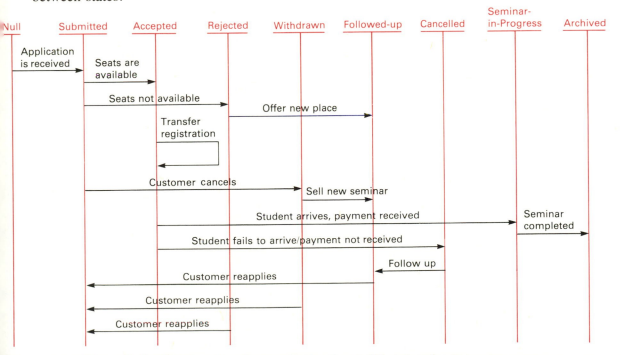

Figure 18.10   State-transition diagram showing the possible states of a registration for a seminar.

With this type of diagram, the analyst can draw the *life cycle* of an entity type, showing what states it can be in and what transitions can occur. The information on Fig. 18.10 can be drawn in a table like Fig. 18.5 to help ensure complete representation of possible state transitions.

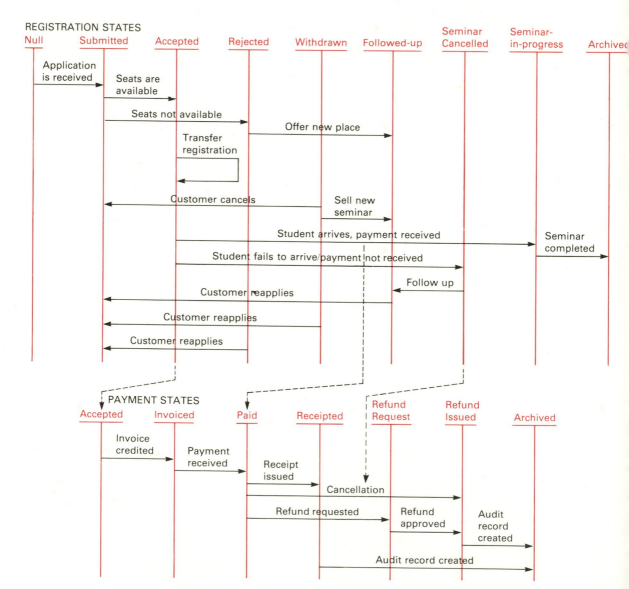

Figure 18.11 Payment states and registration states with linkages between them. One state-transition diagram must relate to mutually exclusive states.

NON-MUTUALLY EXCLUSIVE STATES

In a finite-state-machine diagram, the states represented *must be mutually exclusive*. The machine cannot be in two of the set of states at the same time.

Often an entity type can be in states which are not mutually exclusive. The registration in Fig. 18.10, for example, might be *paid* or *not paid*. The state of payment cannot be included among the states of Fig. 18.10 because it is not mutually exclusive with them. A *not-paid* registration could be in the states *accepted, rejected, seminar-in-progress*, and so on. A *separate* state-transition diagram must, therefore, be drawn for the payment states.

The separate payment-states diagram is shown at the bottom of Fig. 18.11. It shows the life cycle of payments for the seminar registration. There are linkages between the registration-states diagram and the payment-states diagram. The *accepted* state on the registration-states diagram corresponds to the *accepted* state on the payment-states diagram. The *paid* state on the payment-states diagram is an input condition necessary before allowing the student to take the seminar which is shown on the registration-states diagram. The *canceled* state on the upper diagram is an input to the *refund-issued* state on the lower diagram.

MUTUALLY EXCLUSIVE TRANSITIONS

Whereas states *must* be mutually exclusive, transitions between states are only sometimes mutually exclusive. One type of transition *or* another may occur. A choice of transitions may be drawn with a branching line with a dot at the branch—the convention we have used on other diagrams:

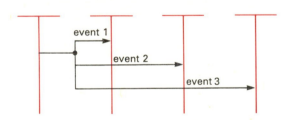

For example, in Fig. 18.11, after an application is received, there may or may not be seats available. A transfer therefore occurs from SUBMITTED to either ACCEPTED or REJECTED. This can be drawn as follows:

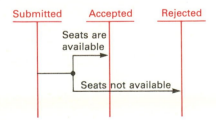

WHEN SHOULD STATE-TRANSITION DIAGRAMS BE USED?

State-transition diagrams are a useful tool in the systems analyst's kit. They are easy to draw and help to clarify situations where multiple changes of state occur. Like decision tables, they are not needed in the design of every type of system, but for certain types of systems, they are invaluable.

They are useful in showing the multiple states possible for entity types in data-base systems. They are useful for diagramming the behavior of systems with multiple message types with complex processing and synchronization requirements. Fence diagrams of state transitions are particularly useful for systems with many transitions among a relatively small number of states. The table of state transitions (like Fig. 18.6) encourages the analyst to examine all possible state transitions. He is less likely to forget one that might be critical.

REFERENCES

1. *Recommendation X.25: Interface Between Data Terminal Equipment and Data Circuit-Termination Equipment for Terminals Operated in the Packet Mode on Public Data Networks*, CCITT, Geneva, 1977.

2. T. Pietkowski introduced this form of diagram and finite-state techniques to specify IBM's Systems Network Architecture.

3. *IBM Systems Network Architecture, Format and Protocol Reference Manual: Architecture Logic*, Manual SC 30.3112, IBM, White Plains, NY, 1977.

19 DATA STRUCTURE DIAGRAMS

INTRODUCTION   Many structured techniques represent data as tree structures (hierarchies). This is illustrated for Michael Jackson methodology in Fig. 12.1 and for Warnier–Orr methodology in Fig. 11.2.

The data on purchase orders, bank statements, restaurant menus, and most computer printouts *can* be represented as tree structures. However, some computer input and output data are not hierarchical. Such data can be drawn as tree structures only if certain data items are shown redundantly or if there are non-tree-structured associations linking the trees. Nonhierarchical data structures are referred to as *plex* or *network* structures.

Plex (network) structured data are extremely important in the data-base environment. This chapter and Chapter 20 illustrate the ways in which data-base structures are drawn. This chapter illustrates:

- Bubble charts
- Drawings of record structures

Chapter 20 illustrates:

- Entity-relationship diagrams
- Data models

BUBBLE CHARTS   Bubble charts provide a way of drawing and understanding the associations among data items. This understanding is necessary in order to create records which are clearly structured. Bubble charts form the input to a data modeling process which creates stable data structures. This process is automated.

Bubble charts are a useful way to teach end users and analysts about associations in data. When they start to create logical data-base structures they should employ bubble charts. As they become more expert they may avoid drawing them and represent the same information as input to an automated tool which synthesizes the data structure [1].

The most elemental piece of data is called a *data item*. It is sometimes also called a *field* or a *data element*. It is the atom of data, in that it cannot be subdivided into smaller data types and retain any meaning to the users of the data. You cannot split the data item called SALARY, for example, into smaller data items which by themselves are meaningful to end users.

In a bubble chart each *type* of data item is drawn as an ellipse:

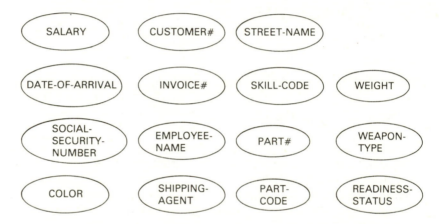

A data base contains hundreds (and sometimes thousands) of types of data items. Several thousand types of data items may be used in the running of a big corporation.

In order to computerize the activities of a corporation, the data items it uses must be defined, cataloged, and organized. This is often difficult and time consuming because data have been treated rather sloppily in the past. What is essentially the same data item type has been defined differently in different places, represented differently in computers, and given different names. Data item types which were casually thought to be the same are found to be not quite the same.

The data administrator has the job of cleaning up this confusion. Definitions of data item types must be agreed upon and documented. Much help from end users is often needed in this process.

ASSOCIATIONS BETWEEN DATA ITEMS

A data item by itself is not of much use. For example, a value of SALARY by itself is uninteresting. It only becomes interesting when it is associated with another data item, such as EMPLOYEE-NAME, thus:

A data base, therefore, consists not only of data items but also of associations among them. There are a large number of different data-item types and we need a map showing how they are associated. This map is sometimes called a *data model*.

ONE-TO-ONE AND ONE-TO-MANY ASSOCIATIONS

There are two kinds of links that we shall draw between data items: a one-to-one association and a one-to-many association.

A *one-to-one association* from data-item type A to data-item type B means that *at each instant in time, each value of A has one and only one value of B associated with it*. There is a one-to-one mapping from A to B. If you know the value of A, you can know the value of B.

There is only one value of SALARY associated with any value of EMPLOYEE# at one instant in time; therefore, we can draw a one-to-one link from EMPLOYEE# to SALARY. It is drawn as a small bar across the link, thus:

The reader may think of the bar as being a "1." It is said that EMPLOYEE# *identifies* SALARY. If you know the value of EMPLOYEE#, you can know the value of SALARY.

A *one-to-many link* from A to B means that *one value of A has one or many values of B associated with it*. This is drawn with a crow's-foot.

Whereas an employee can have only one salary at a given time, he might have one or many girl friends. Therefore, we would draw

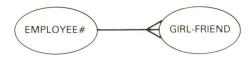

For one value of the data-item type EMPLOYEE# there can be one or many values of the data-item type GIRL-FRIEND.

We can draw both of the foregoing situations on one bubble chart, thus:

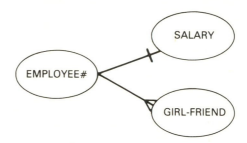

This bubble chart *synthesizes* the two preceding charts into one chart. From this one chart we could derive either of the two previous charts.

The two previous charts might be two different user views, one user being interested in salary and the other in girl friends. We have created one simple data structure which incorporates these two user views. This is what the data administrator does when building a data base, but the real-life user views are much more complicated than the illustration above and there are many of them. The resulting data model sometimes has hundreds or even thousands of data-item types.

Note: Many analysts draw the one-to-one and one-to-many associations as single-headed and double-headed arrows, thus:

The senior author has used single-headed and double-headed arrows in earlier books, but they are avoided here because arrows tend to suggest a flow or time sequence, and are used extensively for this purpose in other types of diagrams.

The one-to-one symbol is extremely important in the processes of normalizing and synthesizing data—the basis of data analysis.

TYPES AND INSTANCES   The terms with which we describe data can refer to *types* of data or to *instances* of that data. "EMPLOYEE-NAME" refers to a type of data item. "FRED SMITH" is an instance of this data-item type. "EMPLOYEE" may refer to a type of record. There are many instances of this record type, one for each person employed. The diagrams in this chapter show *types* of data, not instances. A data model shows the associations among *types* of data.

The bubble chart shows data-item types. There are many occurrences of each data-item type. In the example above there are many employees, each with a salary and with zero, one, or many girl friends. The reader might imagine a third dimension to the bubble charts showing the many values of each data-item type, thus:

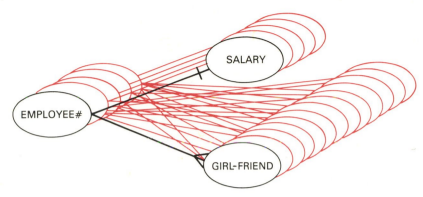

In discussing data we ought to distinguish between types and instances. Sometimes abbreviated wording is used in literature about data. The words "DATA ITEM" or "RECORD" are used to mean "DATA-ITEM TYPE" or "RECORD TYPE."

REVERSE ASSOCIATIONS

Between any two data-item types there can be a mapping in both directions. This gives four possibilities for forward and reverse association. If the data-item types are MAN and WOMAN, and the relationship between them represents *marriage*, the four theoretical possibilities are:

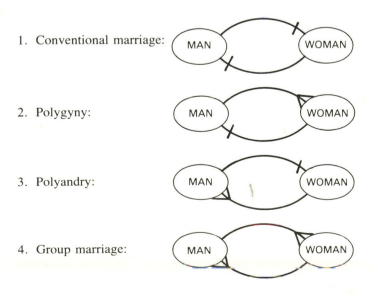

1. Conventional marriage:

2. Polygyny:

3. Polyandry:

4. Group marriage:

The reverse associations are not always of interest. For example, with the bubble chart

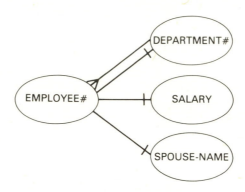

we want the reverse association from DEPARTMENT# to EMPLOYEE# because users want to know what employees work in a given department. However, there is no link from SPOUSE-NAME to EMPLOYEE# because no user wants to ask "What employee has a spouse named Gertrude?" If a user wanted to ask "What employees have a salary over $25,000?", we might include a crow's-foot link from SALARY to EMPLOYEE#.

OPTIONAL DATA ITEMS

Sometimes there may be zero or one data item associated with another. This is indicated by putting a "small circle" on the link, by the "one" or "many" indicator, thus:

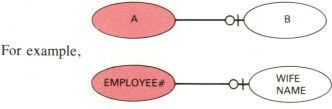

For example,

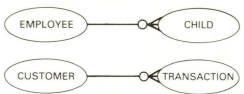

Optionality applies to one-to-many associations also:

![EMPLOYEE to CHILD and CUSTOMER to TRANSACTION diagram]

If the optional bubble is a nonprime attribute (rather than a primary key), it may be treated like any other nonprime attribute when synthesizing the data model.

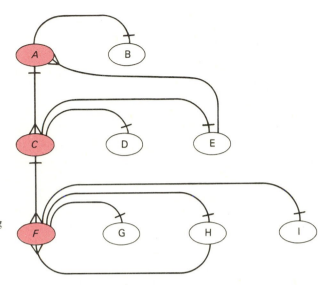

Figure 19.1 Bubble chart showing one-to-one (———+—) and one-to-many (———≺) associations among data-item types.

KEYS AND ATTRIBUTES

Given the bubble chart method of representing data, we can state three important definitions:

1. Primary key

2. Secondary key

3. Attribute

A *primary key* is a bubble with one or more one-to-one links going to another bubble. Thus, in Fig. 19.1, A, C, and F are primary keys. A primary key may uniquely identify many data items. We will color the primary key *red*.

Data items that are not primary keys are referred to as *nonprime attributes*. All data items, then, are either *primary keys* or *nonprime attributes*.

In Fig. 19.1, B, D, E, G, H, and I are nonprime attributes. Often the word "attribute" is used instead of "nonprime attribute." Strictly, the primary key data items are also attributes. EMPLOYEE# is an attribute of the employee.

The names of data-item types which are primary keys are underlined in the bubble charts and drawings of records. We can define a nonprime attribute as follows: A *nonprime attribute* is a bubble with no one-to-one links going to another bubble. Each primary key uniquely identifies one or more data items. Those which are not other primary keys are attributes.

A secondary key does not uniquely identify another data item. One value of a secondary key is associated with one or many values of another data item. In other words, there is a crow's-foot link going from it to that other item. A *secondary key* is a nonprime attribute with one or more crow's-foot links to another data item. In Fig. 19.1, E and H are secondary keys.

For emphasis the box on p. 238 repeats these three fundamental definitions.

A *primary key* is a bubble with one or more one-to-one links going to another bubble.

A *nonprime attribute* is a bubble with no one-to-one link going to another bubble.

A *secondary key* is an attribute with one or more one-to-many links going to another bubble.

DATA-ITEM GROUPS

When using a data base we need to extract multiple different views of data from one overall data-base structure. The bubble charts representing these different views of data can be merged into one overall chart. In the bubble chart which results from combining many user views, the bubbles are grouped by primary key. Each primary key is the unique identifier of a group of data-item types. It has one-to-one links to each nonprime attribute in that group.

The data-item group needs to be structured carefully so that it is as stable as possible. We should not group together an ad hoc collection of data items. There are formal rules for structuring the data-item group, which are part of the normalization process [1].

RECORDS

The data-item group is commonly called a *record*—sometimes a *logical record*—to distinguish it from whatever may be stored physically. A record is often drawn as a bar containing the names of its data items, as shown in Fig. 19.2.

The record in Fig. 19.2 represents the following bubble chart:

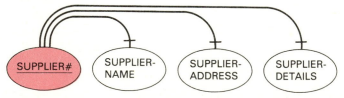

It may be useful to split the SUPPLIER-ADDRESS data item into component data items, thus:

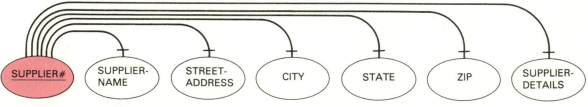

This is useful only if the components may be referenced individually.

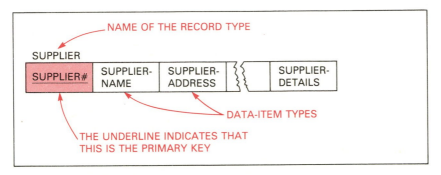

Figure 19.2   Drawing of a record.

Figure 19.3   Record in Fig. 19.2 redrawn to show the decomposition of SUP-PLIER-ADDRESS. These components do not themselves constitute a record or data-item group with a primary key.

Figure 19.3 shows the record redrawn to show that STREET-ADDRESS, CITY, STATE, and ZIP are collectively referred to as SUPPLIER-ADDRESS but are not by themselves a record with a primary key.

CONCATENATED KEYS

Some data-item types cannot be identified by any single data-item type in a user's view. They need a primary key (unique identifier) which is composed of more than one data item type in combination. This is called a *concatenated key*.

Several suppliers may supply a part and each charge a different price for it. The primary key SUPPLIER# is used for identifying information about a *supplier*. The key PART# is used for identifying information about a *part*. Neither of those keys is sufficient for identifying the *price*. The price is dependent on both the supplier and the part. We create a new key to identify the price, which consists of SUPPLIER# and PART# joined together (concatenated). We draw this as one bubble:

The two fields from which the concatenated key is created are joined with a " + " symbol.

The concatenated key has one-to-one links to the keys SUPPLIER# and PART#. The resulting graph is as follows:

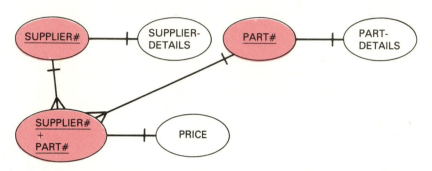

By introducing this form of concatenated key into the logical view of data, we make each data item dependent on one key bubble. Whenever a concatenated key is introduced, the designer should ensure that the items it identifies are dependent on the whole key, not on a portion of it only.

In practice it is sometimes necessary to join together more than two data-item types in a concatenated key. For example, a company supplies a product to domestic and industrial customers. It charges a different price to different *types of customers*, and also the price varies from one *state* to another. There is a *discount* giving different price reductions for different quantities purchased. The *price* is identified by a combination of CUSTOMER-TYPE, STATE, DIS-COUNT, and PRODUCT, thus:

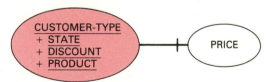

The use of concatenated keys gives each data-item group in the resulting data model a simple structure in which each nonprime attribute is fully dependent on the key bubble and nothing else:

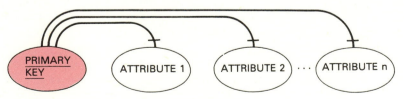

DERIVED DATA   Certain data items are derived by calculation from other data items. For example, TOTAL-AMOUNT on an invoice may be derived by adding the AMOUNT data items on individual lines. A derived data-item type may be marked on a bubble chart by shading its ellipse. Dashed lines or colored lines may be drawn to the derived data-item type from the data-item types from which it is derived. Figure 19.4 illustrates this.

Where possible the calculation for deriving a data item may be written on the diagram, as in Fig. 19.4. Sometimes the computation may be too complex and the diagram refers to a separate specification. It might refer to a decision tree or table such as Fig. 17.2, which shows a derived data item.

Derived data items may or may not be stored with the data. They might be calculated whenever the data are retrieved. To store them requires more storage; to calculate them each time requires more processing. As storage drops in cost it is increasingly attractive to store them. The diagrams initially drawn are *logical* representations of data which represent derived data without saying whether or not the data are stored. This is a later, physical decision.

There has been much debate about whether derived data-item types should be shown on diagrams of data or data models. In the author's view they should be shown. Some fourth-generation or nonprocedural languages cause data to be derived automatically once statements like those in Fig. 19.4 are made describing the derivation.

The shading means that the data item is derived from other data items

X + Y = Z

Z is derived from X and Y

ORDER#	CUSTOMER	ADDRESS	TOTAL

TOTAL = Σ AMOUNT

TOTAL is derived from many AMOUNT data items

ITEM#	DESCRIPTION	AMOUNT

Figure 19.4   Derived data-item types shown on diagrams of data.

DATA ANALYSIS When data analysis is performed, the analyst examines the data-item types that are needed and draws a diagram of the dependencies among the data items. The one-to-one and one-to-many links which we have drawn on bubble charts are also drawn between records, as in Fig. 19.4.

Figures 19.5 to 19.7 illustrate data analysis. Figure 19.5 shows a sales contract. The data-item types on this contract are charted in Fig. 19.6. The bubble chart of Fig. 19.6 is redrawn as a record diagram in Fig. 19.7.

In drawing associations between records it helps to draw the one-to-one indicators pointing up and the one-to-many indicators pointing down, where possible. Lines with a one-to-one indicator at one end and a crow's foot at the

THE HOUSE OF MUSIC INC.
A Collins Corporation
Main Office
108 Old Street, White Cliffs, IL 67309
063 259 0003

SALES CONTRACT

Contract No. 7094

SOLD BY Mike	DATE 6/10/83

Name __Herbert H. Matlock__

Address __1901 Keel Road__

City __Ramsbottom, Illinois__ Zip __64736__

Phone __063 259 3730__ Customer # __18306__

REMARKS:

10 yrs. parts and labor on the Piano
1 yr. parts and labor on pianocorder

Delivery Address:

DESCRIPTION	PRICE	DISCOUNT	AMOUNT
New Samick 5'2" Grand Piano model G-1A			
# 820991 with Marantz P-101 # 11359			9500.00
		TOTAL AMOUNT	9500.00
		TRADE IN ALLOWANCE	2300.00
		SALES TAX	
		DEPOSIT	1000.00
		FINAL BALANCE	6200.00

PLEASE NOTE: All sales pending approval by management and verification of trade-in description.

If this contract is breached by the BUYER, the SELLER may take appropriate legal action, or, at its option, retain the deposit as liquidated damages.

Buyer's Signature _____

Figure 19.5 Sales contract. The data-item types on this document and their associations are diagrammed in Figs. 19.6 and 19.7.

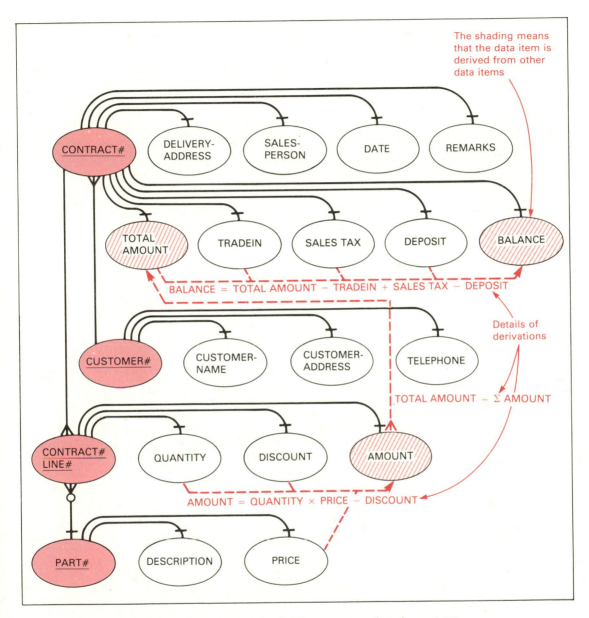

The shading means that the data item is derived from other data items

BALANCE = TOTAL AMOUNT − TRADEIN + SALES TAX − DEPOSIT

Details of derivations

TOTAL AMOUNT − Σ AMOUNT

AMOUNT = QUANTITY × PRICE − DISCOUNT

Figure 19.6   Bubble chart showing the data-item types on the sales contract of Fig. 19.5, and the associations among them.

Note: DESCRIPTION and PRICE are functionally dependent on PART#. QUANTITY, DISCOUNT, and AMOUNT are functionally dependent on CONTRACT# + LINE#.

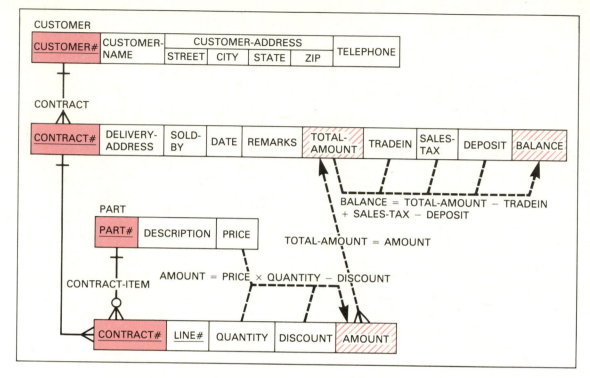

Figure 19.7 Record diagram of the data in Fig. 19.6. The derivation expressions could be recorded separately.

other end are the basic component of hierarchies and can be drawn as follows:

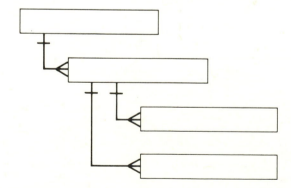

The offsetting to the right in the structure shows the depth in the tree, as discussed in Chapter 3.

The data shown in Figs. 19.6 and 19.7 represent a portion of the data likely to reside in a data base. The data base contains more detail identified by the keys

CUSTOMER# and PART#, for example. An overall data model, drawn something like Fig. 19.7, represents the data in the data base or a conceptual representation of data needed for running data processing which may reside in multiple data bases. This is discussed in Chapter 20.

BOX 19.1   Notation used on bubble charts for data analysis (compare with Box 20.1)

A data-item (field) type is drawn as a named bubble.

1:1 Association

- A identifies B.
- B is functionally dependent on A.
- For one occurrence of A there is always one and only one occurrence of B.

1:M Association

For one occurrence of A there are one or multiple occurrences of B.

Optional 1:1 Association

For one occurrence of A there is zero or one occurrence of B.

Optional 1:M Association

For one occurrence of A there are zero, one, or multiple occurrences of B.

Labeled Associations

A label may be written on an association. This is normally done when two associations with different meanings exist between the same two data-item types.

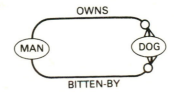

BOX 19.1   *(Continued)*

Grouped Data-Item Types

Several data-item types are
given a group name but do
not constitute a record.

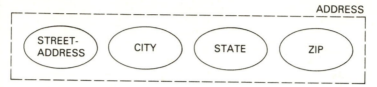

Primary Key

A primary key is a data-item
type with a one-to-one link
to other data-item types.
That is, it *identifies* other
data-item types. The name
of the primary key data-item
is underlined.

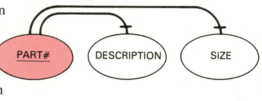

Concatenated Keys

When a primary key consists
of multiple data-item types,
these are drawn as one bubble.
Their names are separated with
a "+".

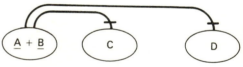

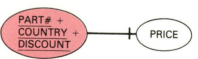

The components of the
concatenated key are drawn
on the chart.

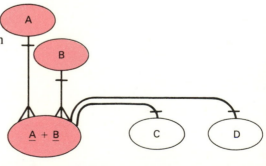

BOX 19.1   *(Continued)*

Derived Data

The shading of a bubble means that
this data item is derived
from other data items.

The boldface arrow shows from which
data-item types a derived
data-item type is obtained.

The derivation equation
may be written by the
arrow.

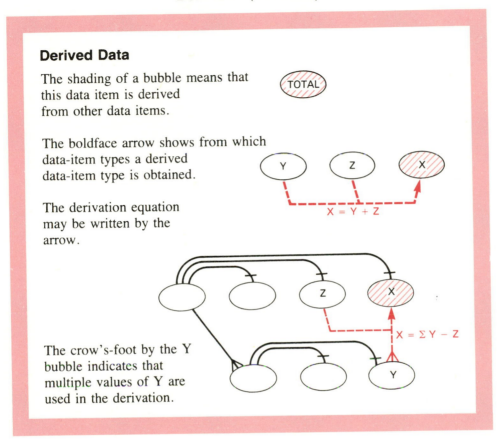

The crow's-foot by the Y
bubble indicates that
multiple values of Y are
used in the derivation.

REFERENCE

1. James Martin, *Managing the Data-Base Environment*, Prentice-Hall, Inc., NJ,
1983.

20 ENTITY-RELATIONSHIP DIAGRAMS

INTRODUCTION Chapter 19 contains detailed low-level diagrams of data. This chapter contains high-level overview diagrams of data which are used in strategic or top-down planning [1].

Top-down planning of data identifies the entity types involved in running an enterprise and determines the relationships among these entity types. An entity-relationship diagram is developed which can be further decomposed into detailed data models.

In order to run an enterprise efficiently certain data are needed. These data are needed regardless of whether computers are used, but computers provide great power in getting the right data to the right people. The data in question need to be planned and described. We need data about these data. Data about data are referred to as *metadata*. A data *model* contains metadata.

Data analysts need much help from end users and user executives to enable them to understand an organization's data, and to design the data that will be most useful in managing the organization. They need clear ways of diagramming the data. Charts like those in this chapter are an essential part of the overall planning of an organization's information resources.

ENTITIES An entity is something (real or abstract) about which we store data. Examples of entity types are CUSTOMER, PART, EMPLOYEE, INVOICE, MACHINE TOOL, SALESPERSON, BRANCH OFFICE, SALES TV AREA, WAREHOUSE, WAREHOUSE BIN, SHOP ORDER, SHIFT REPORT, PRODUCT SPECIFICATION, LEDGER ACCOUNT, JOURNAL POSTING ACCOUNT, PAYMENT, CASH RECEIPT, DEBTOR, CREDITOR, and DEBTOR ANALYSIS RECORD.

The name of each entity type should be a noun, sometimes with a modifier word. An entity type may be thought of as having the properties of a noun. An entity has various *attributes* which we wish to record, such as color, monetary value, percentage utilization, or name.

An *entity type* is a named class of entities which have the same set of *attribute* types; for example, EMPLOYEE is an entity type. An *entity instance* is one specific occurrence of an entity type: for example, B. J. WATKINS is an instance of the entity type EMPLOYEE.

We describe data in terms of entity types and attributes. An entity type has multiple attributes. For example, the entity type PART may have the attributes PART#, NAME, TYPE, COLOR, SIZE, QUANTITY__IN__STOCK and REORDER__QUANTITY.

A rectangular box is drawn to represent an entity type. For most entities we store records: for example, CUSTOMER records, PART records, EM-PLOYEE records, and so on. The box is sometimes also used to show an entity record type. However, the intent of our data model is to represent the reality of the data without yet thinking about how we will represent it in computers. We may decide to represent it without a conventional record structure.

ENTITY CHARTS

On an *entity-relationship diagram* (often called simply an *entity chart*) the boxes are interconnected by links that represent associations between entity types. The data in Fig. 19.7 contain four entities: CUSTOMER, PART, CONTRACT, and CONTRACT ITEM. An entity-relationship diagram can be drawn using the same notation as that used in Chapter 19, thus:

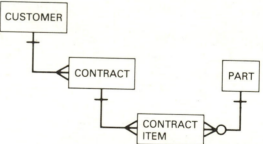

This chart shows that a customer can have multiple contracts. A contract is for one customer and can be for multiple contract items. There are zero, one, or many contracts for each part. A contract item relates to one contract and one part.

As before, a one-to-one association is drawn with a bar across the link:

A one-to-many association is drawn with a crow's-foot:

Where the cardinality may be zero the cardinality symbol includes a ''o'':

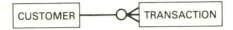

An optional link has a small circle at the start of the link:

Readers should examine Fig. 20.1 to ensure that they understand the meaning of the links.

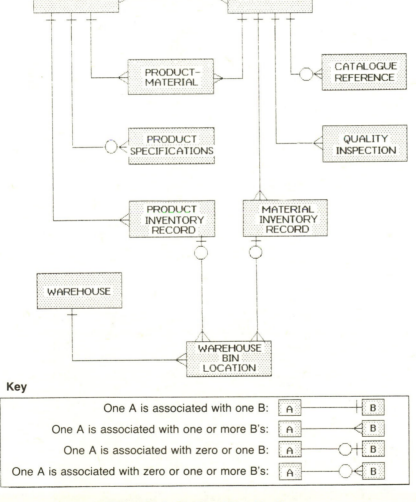

Figure 20.1 Entity-relationship diagram.

CONCATENATED ENTITY TYPE

Some important information does not relate to one entity type alone, but to a conjunction of entity types. For example, Fig. 19.1 has PRODUCT and MATERIAL. We want to record how much of a given material is used on a given product. This requires a concatenated entity type, PRODUCT + MATERIAL. It is not information about the product alone or the material alone:

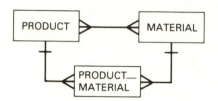

Whenever a link has a crow's-foot at both ends, the designer should ask whether there is any information that would need a concatenation of the two entities. Usually there is.

Mutually Exclusive Associations

Some associations are mutually exclusive. If A can be associated with either B or C, but not with both, we draw

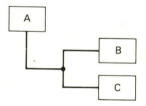

The line branching at a dot is the convention that we have seen on other types of diagrams to show mutual exclusivity. Suppose, for example, that an aircraft is permitted (for regulatory reasons, for example) to carry cargo or passengers but not both. We can draw the following:

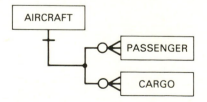

If a driver can be allocated to a truck, a car, or a motor bike, but only to one of these, we draw

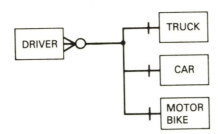

Mutually Inclusive Associations

Some associations are mutually *inclusive*. In other words, if A is associated with B, it must also be associated with C. We draw this with a line branching (without a dot), as on other types of diagrams:

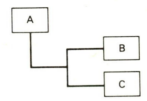

This is rarely used in practice; it would usually be satisfactory to show a mandatory link between B and C.

Subset Associations

One association may be a subset of another. In this case, the associations are linked by a line with an "S" around the subset association, thus:

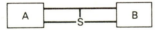

Here the bottom association is a subset of the top one.

Labels and Sentences

The link between entities should be thought of as forming a simple sentence. The entity at the start of the link is the subject of the sentence and the entity which the link goes to is the object.

For the following link:

the sentence is

AN INVOICE HAS ONE OR MULTIPLE LINE_ITEMS.

For the following labeled link connecting a passenger record to a special-catering record:

the sentence is

PASSENGER REQUIRES ZERO OR ONE SPECIAL___CATERING.

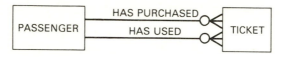

In this case, the sentences formed are

PASSENGER HAS PURCHASED ZERO OR ONE TICKET(S)
PASSENGER HAS USED ZERO OR ONE TICKET(S)

Links in Both Directions

Between two entity types there can be associations in both directions. We *could* draw this with two links:

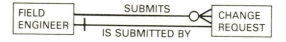

Usually, however, the information about both directions is combined on to a single link:

Thus there are two sentences associated with one bidirectional link:

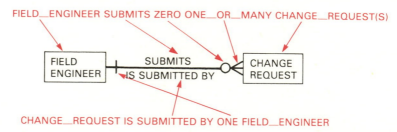

Sometimes the two sentences have effectively the same meaning. In other cases, they have entirely different meanings.

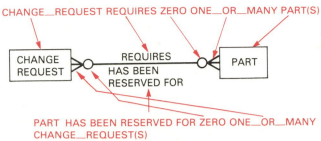

CHANGE_REQUEST REQUIRES ZERO ONE_OR_MANY PART(S)

PART HAS BEEN RESERVED FOR ZERO ONE_OR_MANY CHANGE_REQUEST(S)

In order to be clear regarding which label applies to which direction of the link, the links are drawn horizontally and vertically and the following convention applies. On vertical links the left-hand label is read going down and the right-hand label is read going up. On horizontal links, the upper label is read going from left to right, and the lower label is read going from right to left, thus:

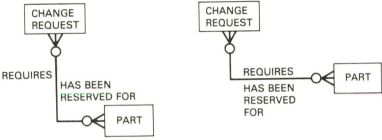

Sentence building, as described above, should be enforced when links are labeled. A label that does not form a sentence can be vague, thus:

Some data analysts like to label every link. This takes time, and the additional work is often not worthwhile. The meanings of most links are obvious. Where a link could have alternative meanings, it should be labeled. For example, it may be important to state that a LOADSHEET records *on-board* PASSENGERS and BAGGAGE, not merely passengers who are booked or who have checked in:

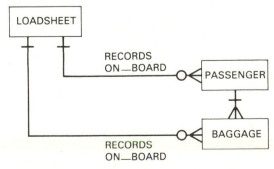

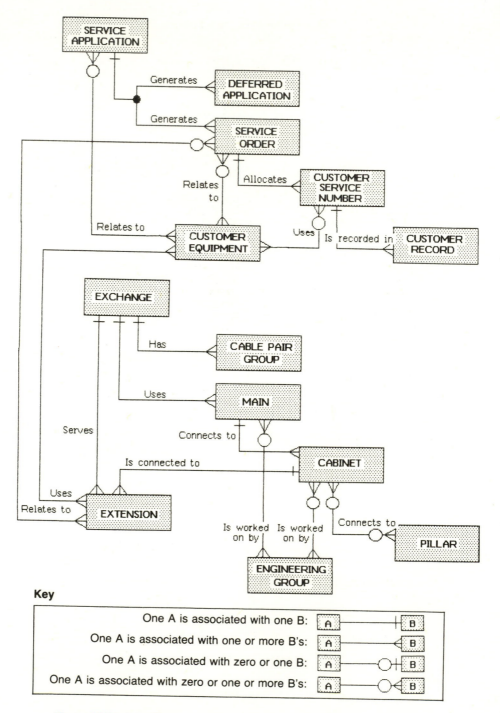

Key

One A is associated with one B:	A ⊢——⊢⊢ B
One A is associated with one or more B's:	A ⊢——⊰ B
One A is associated with zero or one B:	A ⊢——○⊣⊢ B
One A is associated with zero or one or more B's:	A ⊢——○⊰ B

Figure 20.2 Portion of an entity-relationship diagram for a telephone company, showing labeled associations.

Fig. 20.2 shows an entity chart for a telephone company with labeled links.

SUBJECT AND PREDICATE

The information in an information system can be thought of as consisting of statements—factual assertions on topics of concern to the enterprise. For example:

- B. J. Watkins manages the Sales Department.
- K. L. Jones works for the Sales Department.
- March had a net-after-tax-profit of $150,000
- Order #72193 has a due-date of June 17.

Sentences can be passed into a subject and predicate. The predicate can be decomposed into a descriptor and an association which connects the descriptor to the subject:

	PREDICATE	
SUBJECT	ASSOCIATION	DESCRIPTOR
B. J. Watkins	manages	the Sales Department
K. L. Jones	works for	the Sales Department
March	had a net-after-tax-profit of	$150,000
Order # 72193	has a due-date of	June 17

The components of the sentences above are entities, links, attributes, or attribute values:

	PREDICATE	
SUBJECT	ASSOCIATION	DESCRIPTOR
B. J. Watkins (ENTITY)	manages (LINK)	the Sales Department (ENTITY)
K. L. Jones (ENTITY)	works for (LINK)	the Sales Department (ENTITY)
March (ENTITY)	had a net-after-tax-profit of (ATTRIBUTE)	$150,000 (VALUE)
Order # 72193 (ENTITY)	has a due-date of (ATTRIBUTE)	June 17 (VALUE)

A data model is a framework into which multiple values can be recorded. The values change while the framework remains the same. The flight information display at an airport is a framework like a simple data model into which changing values are placed.

We may draw the framework for the foregoing sentences showing the entity types as rectangles. Attributes can be drawn as ellipses (as in Chapter 19). The frameworks or models for the four sentences above are as follows:

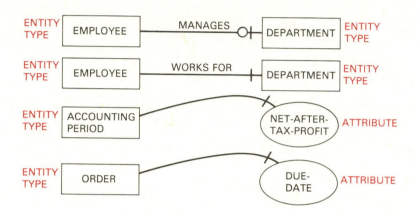

Some research work is in progress on advanced forms of data models where the meaning of the link is encoded so as to be intelligible to computers. This can be used in advanced languages that employ the data base.

BASIC CONSTRUCTS

It will be seen that there are four basic constructs in representing data: entity type, attribute, value, and link. Sometimes a more subtle construct is needed, a relationship between links, of which mutual exclusivity and mutual inclusivity are examples. We may also use formulas to show how derived data are obtained.

An entity-relationship chart contains *entity types* and *links*. Sometimes link connectors are used. As the entity chart is expanded into a *full* data model, *attributes* are added, and sometimes formulas for derived attributes. When the model is translated into implementations, the attributes assume *attribute values*.

Integrity checks can be stated on links and attributes. A link can have a maximum and minimum value of cardinality (how many of entity B can be associated with entity A). An attribute may have a maximum and a minimum value, or a discrete set of values. An entity value may be constrained to be of a stated data type.

Box 20.1 summarizes these constructs.

BOX 20.1   Basic constructs used for describing data

ON A DETAILED DATA MODEL

ON AN ENTITY-RELATIONSHIP CHART

- ENTITY TYPE
  An entity is something about which we store data.
  An entity type refers to a class of entities about which the same attributes are kept.
- LINK
  An association between two entity types showing how they are related.
- LINK CONNECTOR
  A connection between two or more links showing how they are related.

- ATTRIBUTE
  A single piece of information about an entity type.
- FORMULA FOR DERIVED ATTRIBUTE
  A means of computing the value of an attribute which is derived from other attributes.

- ATTRIBUTE VALUE
  A symbol denoting some quality or quantity which is used to describe entities, and which is used as an instance of an attribute.

LOOPED ASSOCIATIONS

Sometimes an occurrence of an entity of a given type is associated with other occurrences of the same type. In this case we draw a loop, thus:

For example, in a zoo data base we might wish to record which animals are children of other animals and which animal is the mother:

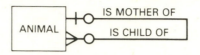

It is normally desirable to label looped associations. Sometimes more than one loop is necessary for the same entity type:

Loops are common in a factory bill of materials. A subassembly is composed of other subassemblies.

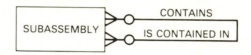

In a personnel data base some employees manage other employees:

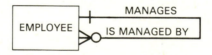

SEMANTIC INDEPENDENCE

The objective of building entity-relationship diagrams and data models is to create a description of the semantics of data which reflects the real enterprise and its informational requirements. The task of the data modeler is to capture reality and communicate about it accurately. The modeler tends to be distracted from this task if he has to think about computer hardware, data-base software, or if the line between the semantics and implementation of data becomes blurred.

A well-structured model keeps one fact in one place (a principle of data normalization). Each semantic building block is intended to be as independent as possible of the others. This has the following practical payoffs [2]:

- Each construct has but one meaning, and each meaning is captured in just one construct. Exceptions and special cases are minimized. Building, reading, learning, and understanding data models is easier and less error-prone.

- The decisions that a data modeler must make become more distinct and independent, so the analyst can deal with them one at a time. Modeling decisions do not have hidden, unexpected consequences.

- Changes in a logical data model are localized. When some aspect of reality changes, only the constructs that directly represent that aspect need to change. To be stable is the prime virtue for a data model, and in a changing world the best stability often is the ability to change gracefully.

- Well-defined and decoupled primitives are the best building blocks for creating complex structures to represent complex realities, because they can be freely combined into structures whose meaning is clear.

- The modelers and users are free to attend to what they know best—the reality of their enterprise and the information they need to know about it.

The clean separation of the semantics of data from other considerations is referred to as *semantic independence*.

INVERTED L CHARTS

In a data model designed to be as useful as it can be, a substantial amount of information needs to be recorded about entities, attributes, and links. Diagrams such as Figs. 20.1 and 20.2 do not have enough space to show all that is needed. The information can be stored in a data dictionary or "encyclopedia," but it also helps to show it graphically.

An entity can be shown as an inverted "L," as shown in Fig. 20.3. Fig. 20.3 shows a subset of a data model with two entities and three links between them. "UI" indicates which attribute or attributes are unique identifiers of the entity. Names of entity types and attributes appear in capital letters. Comments or a description may be written against the top part of the L. The vertical part of the L leaves room for listing many attributes and links.

The *cardinality* of a link refers to the number of entities to which it points. It may point to zero, one, or many. We may be able to make a cardinality statement more precise than "many": for example, <30. In a few cases an exact number can be stated. A person has exactly two parents. A fiscal year has 12 accounting periods. Cardinality limits are written on the link.

Details about the value of an attribute may be written by its name. Fig. 20.3 shows the type of value. A PICTURE clause could be written, showing its format. Range limits or other integrity checks may be shown.

Fig. 20.3 shows attributes and links against the vertical part of the L. Cross-link associations, and complex associations involving multiple attributes or links, may also be shown.

COMPUTER REPRESENTATION OF THE CHART

Entity diagrams can be drawn in a manageable fashion for a dozen or so entities. With hundreds of entities the diagram would be difficult to draw and a mess to maintain without computer graphics.

A computerized tool can identify small hierarchies within a more complex structure and assign levels to the structure as illustrated in Figs. 3.17 to 3.20. Fig.

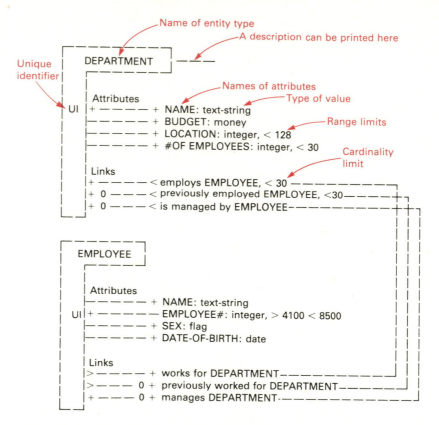

Figure 20.3   Subset of a data model showing two entity types with three links between them. Names of entity types and attributes are shown in capital letters. The diagram is designed for printing on a low-cost character printer.

20.4 shows a complex entity chart simplified by drawing the subhierarchies as in Fig. 3.20. The chart is split into subject areas.

   With computer graphics the chart can be constantly edited, added to, and adjusted in the same way that we adjust text with a word processor. A good computer graphics tool makes change easy; hand-drawn charts make change difficult. Hand-drawn charts of great complexity discourage modification. Often, analysts will do anything to avoid redrawing the chart. This is a serious concern because on complex projects the more interaction, discussion, and modification there are at the planning and design stages, the better the results.

   A chart with hundreds of entities, whether computer drawn or not, is impressive but of little use, except perhaps for the data administrator to hang on his wall in the hope of impressing people. However, with computer graphics small subset charts can be extracted, and these are extremely useful. Subset charts relating to specific data subjects are extracted for checking by end users. Subset charts are extracted for analysts for specific projects. On these subset charts analysts can create data access maps and design data-base procedures. Physical data-base designers employ subset charts for designing data bases. Individual data bases normally employ only a portion of the data represented in a large entity chart.

Figure 20.5 shows an entity diagram of typical complexity drawn with a computerized tool. In this case the tool enables the chart to be changed and added to easily. It does not level the chart, producing subhierarchies as in Fig. 20.4. It does not permit subsets to be extracted automatically for users and analysts. Both of these latter properties are desirable. Figure 20.5 is a computerized COW chart. Computers should be used to clarify the structure and help in employing its information. We regard Fig. 20.5, then, as an example of how computers should *not* be used.

The complete entity chart should be kept in computer form, where it can be conveniently updated and manipulated. From it, small subset charts should be creatable graphically when the data administrator, end users, or systems analysts need to study them, argue about them, and overdraw access maps on them. The entity chart should feed the more detailed data modeling process that follows.

ENTITY SUBTYPES

It is sometimes necessary to divide entity types into entity subtypes. In a zoo, for example, the entity type CREATURE might be subdivided into ANIMAL, FISH, and BIRD. We regard these as entity subtypes *if they have different associations to other entity types.* If, on the other hand, we store essentially the same information about animals, fishes, and birds, we would regard these three categories as merely attribute values of the entity type CREATURE.

We can draw entity subtypes as divisions of the entity-type box, thus:

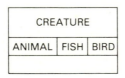

The entity type SATELLITE might be subdivided into LOW-ORBIT and GEOSYNCHRONOUS. LOW-ORBIT SATELLITE has a one-to-many association to the entity type ORBIT DETAIL. GEOSYNCHRONOUS SATELLITE has a one-to-many association with POSITION DETAIL. These two entity types have different attributes. This is drawn as follows on page 268:

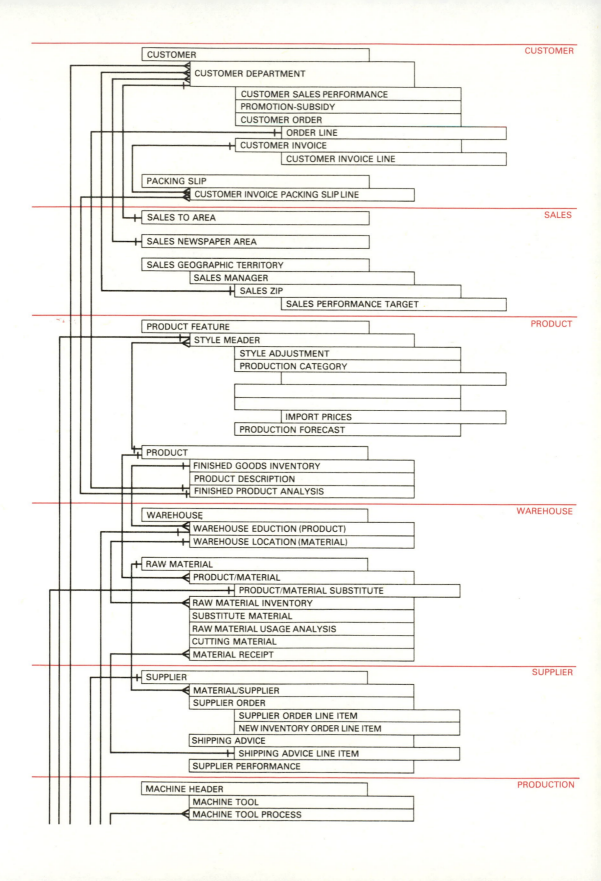

CUSTOMER

CUSTOMER
CUSTOMER DEPARTMENT
CUSTOMER SALES PERFORMANCE
PROMOTION-SUBSIDY
CUSTOMER ORDER
ORDER LINE
CUSTOMER INVOICE
CUSTOMER INVOICE LINE
PACKING SLIP
CUSTOMER INVOICE PACKING SLIP LINE

SALES

SALES TO AREA
SALES NEWSPAPER AREA
SALES GEOGRAPHIC TERRITORY
SALES MANAGER
SALES ZIP
SALES PERFORMANCE TARGET

PRODUCT

PRODUCT FEATURE
STYLE MEADER
STYLE ADJUSTMENT
PRODUCTION CATEGORY

IMPORT PRICES
PRODUCTION FORECAST
PRODUCT
FINISHED GOODS INVENTORY
PRODUCT DESCRIPTION
FINISHED PRODUCT ANALYSIS

WAREHOUSE

WAREHOUSE
WAREHOUSE EDUCTION (PRODUCT)
WAREHOUSE LOCATION (MATERIAL)
RAW MATERIAL
PRODUCT/MATERIAL
PRODUCT/MATERIAL SUBSTITUTE
RAW MATERIAL INVENTORY
SUBSTITUTE MATERIAL
RAW MATERIAL USAGE ANALYSIS
CUTTING MATERIAL
MATERIAL RECEIPT

SUPPLIER

SUPPLIER
MATERIAL/SUPPLIER
SUPPLIER ORDER
SUPPLIER ORDER LINE ITEM
NEW INVENTORY ORDER LINE ITEM
SHIPPING ADVICE
SHIPPING ADVICE LINE ITEM
SUPPLIER PERFORMANCE

PRODUCTION

MACHINE HEADER
MACHINE TOOL
MACHINE TOOL PROCESS

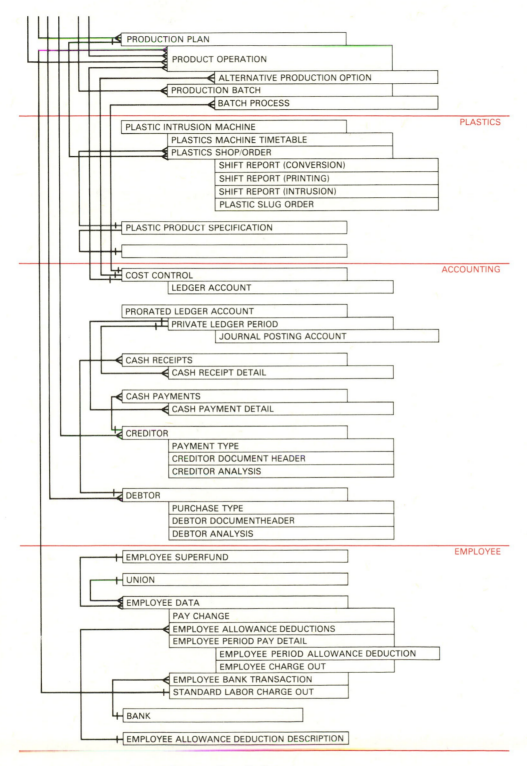

Figure 20.4 Entity diagram for a small textile firm.

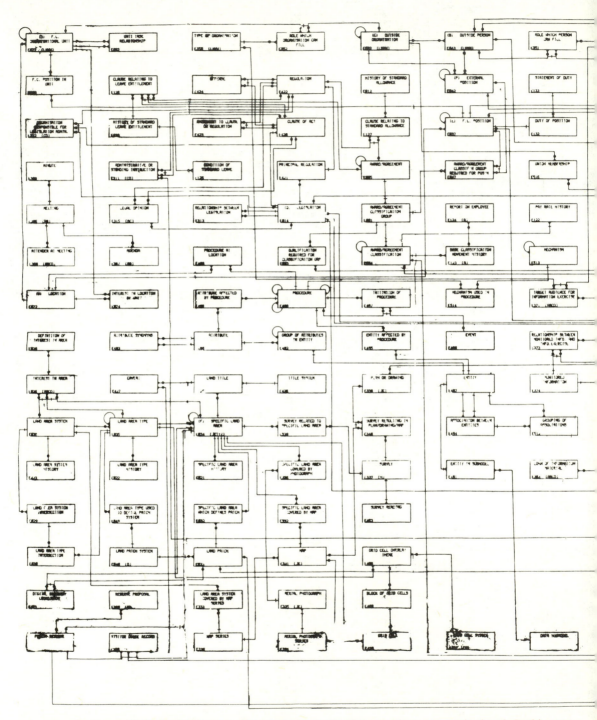

Figure 20.5 Entity diagram of typical complexity drawn with a computerized tool. In this case the tool enables the chart to be changed and added to easily. It does not level the diagram, producing subhierarchies as in Fig. 20.4. It does not permit subsets to be extracted automatically for users

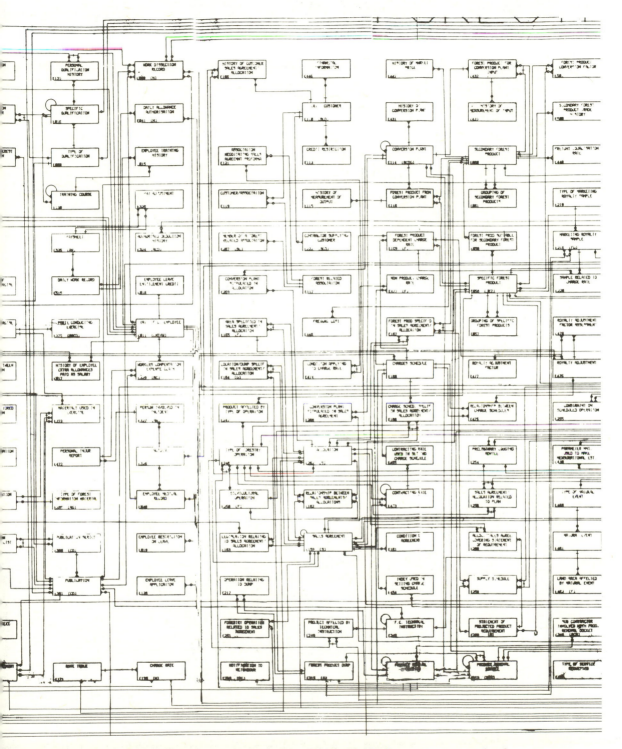

and analysts. Both of the latter properties are desirable. We regard this diagram as an example of how computerized graphics should *not* be used.

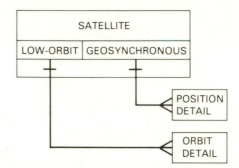

An entity subtype is any subset of entities of a specific entity type about which we wish to record information special to that subtype.

The values of one or more attributes are used to determine the subtype to which a specific entity belongs. These attributes are called the *classifying attributes*.

MULTIPLE SUBTYPE GROUPINGS

The examples above have one category of subtyping which is drawn as a horizontal band in the entity-type box. There may be more than one independent category of subtyping. We then draw multiple bands in the entity-type box. The horizontal bands represent independent subtype groupings.

For example, SATELLITE may be subtyped into MILITARY and CIVIL-IAN, independently of whether it is LOW-ORBIT or GEOSYNCHRONOUS. Again, we store *different* types of information about MILITARY satellites and CIVILIAN satellites.

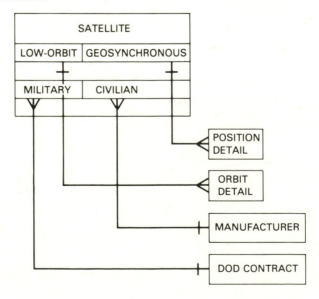

In the illustrations above each horizontal band contains two *mutually exclusive* entity subtypes. Often the entity subtypes are not mutually exclusive. There may be other entities which do not fit into the subtypes shown. This is indicated by leaving a blank space in the horizontal band:

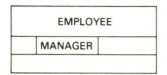 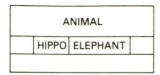

An entity type may contain both mutually exclusive and non-mutually exclusive groupings:

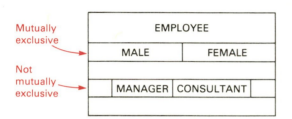 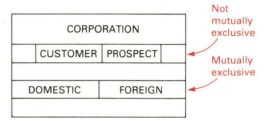

SUBTYPES HIERARCHIES

An entity subtype may itself be subdivided into sub-subtypes:

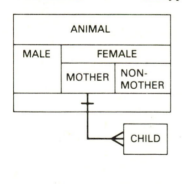 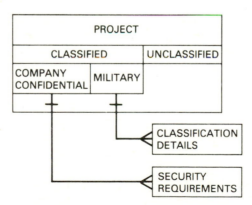

Entity subtypes behave in every way as though they were entity types. They have attributes, and associations to other entities.

Palmer stresses that in his experience of data analysis the newcomer confuses the concepts of entity subtypes, and associations among entity types [3]. He emphasizes the importance of recognizing these to be completely different concepts in spite of the fact that most data-base management systems ignore the concept of entity subtypes.

A similar test can help avoid confusion. We ask "Is A a B?" and "Is B an A?" The permissible answers are ALWAYS, SOMETIMES, and NEVER. If both answers are NEVER, we are not concerned with subtyping. If both answers are ALWAYS, then A and B are synonyms. If the answers are:

"Is A a B?" ALWAYS

"Is B an A?" SOMETIMES

then A is a subtype of B.

Let us look at a case that might be confusing. A somewhat bureaucratic organization has people with the following titles: OFFICIAL, ADVISOR, SUB-AGENT, and REPRESENTATIVE. Should each of these be a separate entity type, or are they subtypes or merely attributes?

The cells in the following table answer the question "Is A a B?"

	B:			
		OFFICIAL	ADVISER	SUBAGENT	REPRESENTATIVE
A: OFFICIAL			Sometimes	Never	Always
ADVISOR		Never		Never	Never
SUBAGENT		Never	Never		Always
REPRESENTATIVE		Sometimes	Never	Sometimes	

The word "always" appears twice. An OFFICIAL and a SUBAGENT are *always* a REPRESENTATIVE. These can be subtypes of the entity type REP-RESENTATIVE. An OFFICIAL is *never* a SUBAGENT, and vice versa, so they are mutually exclusive subtypes. Can there be representatives other than OFFI-CIALS and SUBAGENTS? *Yes*. Therefore, we draw

REPRESENTATIVE		
OFFICIAL	SUBAGENT	

An ADVISOR is *never* any of the others, so that is a separate entity type.

Do we *really* want to regard an OFFICIAL and a SUBAGENT as an entity subtype, or should they be attributes of REPRESENTATIVE? To answer this, we ask "Do they have associations which are different from those of REPRE-SENTATIVE which we need to include in the data model?" *Yes*, they do. An OFFICIAL supervises a SUBAGENT. A SUBAGENT is an external employee working for a CORPORATION about which separate records are kept. An as-sociation from OFFICIAL to SUBAGENT is needed. This can be drawn inside the entity-type box.

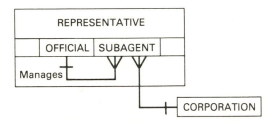

This conveys more information than an attempt at entity analysis without subtyping, which might show the following:

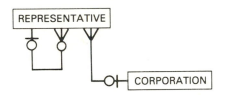

ALTERNATIVE NOTATIONS

Three notations are used for drawing entity diagrams: crow's-foot notation, arrow notation, and Bachman notation. This chapter has used crow's-foot notation. The following table shows the equivalencies among these notations.

	CROW'S-FOOT NOTATION	ARROW NOTATION	BACHMAN NOTATION
A is always associated with one of B	A ⊢⊨ B	A ⟶ B	A ⟶ B
A is always associated with one or many of B	A ⊢< B	A ⊢⊨» B	A ⟶ B
A is associated with zero or one of B	A ⟜○⊦ B	A ⟶○» B	
A is associated with zero, one, or many of B	A ⟜○< B	A ⟶» B	

It is generally desirable to use the same notation for data analysis (Chapter 18) as for entity-relationship diagrams.

Bachman notation was an early notation for drawing data-base structures which deserves a place in the history of systems analysis because it was the first form of data-base diagramming. Bachman notation uses a single-headed arrow for a one-to-many association. It uses an unmarked line for a one-to-one association.

We use an unmarked line to mean that we are uninterested in an association or that we do not yet know what it is. For example, the tail of most links to attributes is unmarked:

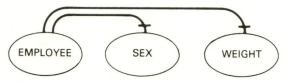

Bachman notation does not distinguish between the uninteresting association and the one-to-one association. It has no specific way to draw functional dependencies—the single most important input to the detailed data modeling process. It cannot draw many of the situations in Box 20.2.

Crow's-foot notation, as far as we can ascertain, was first used and popularized by Ian Palmer. We have used it instead of arrow notation because arrows tend to suggest flows or time sequence. The notation of this and Chapter 19 allows us to achieve consistency among the various diagramming techniques that an analyst needs, as summarized in Chapter 24.

BOX 20.2   Notation used on entity-relationship diagrams

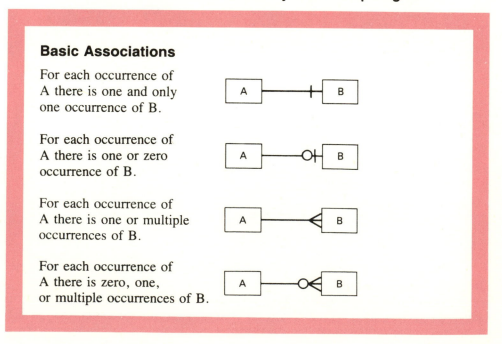

Basic Associations

For each occurrence of
A there is one and only
one occurrence of B.

For each occurrence of
A there is one or zero
occurrence of B.

For each occurrence of
A there is one or multiple
occurrences of B.

For each occurrence of
A there is zero, one,
or multiple occurrences of B.

(Continued)

BOX 20.2 *(continued)*

The relationship of A to B is uninteresting. (This is the tail end of a connector whose head end is necessary.)

The relationship of A to B is of interest but is as yet unknown.

Labeled Associations

A label may be written on the left side of a link going *down:*

On the right side of a link going *up:*

On the top of a link going left to right:

On the bottom of a link going right to left:

Two complementary labels may be written on the same link:

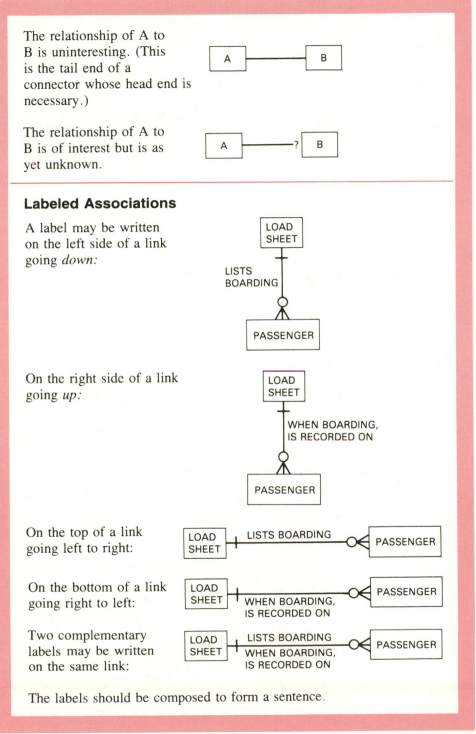

The labels should be composed to form a sentence.

(Continued)

BOX 20.2   *(continued)*

Looped Associations

Any of the associations above may be used in a loop. Here an occurrence of an entity is associated with one or more occurrences of entities of the same type.

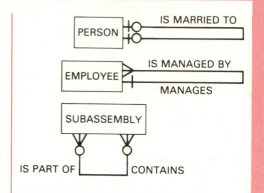

Linked Associations

1. Mutually exclusive associations

   Associations branching from a dot are mutually exclusive. Only one of them can exist for any one occurrence.

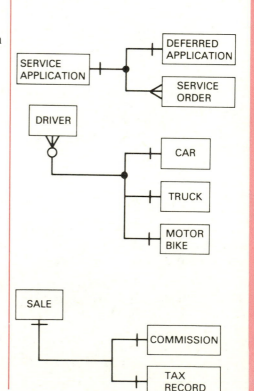

2. Mutually inclusive associations

   Associations connected by a branch are mutually inclusive. All must exist together for any one occurrence.

(Continued)

BOX 20.2   *(continued)*

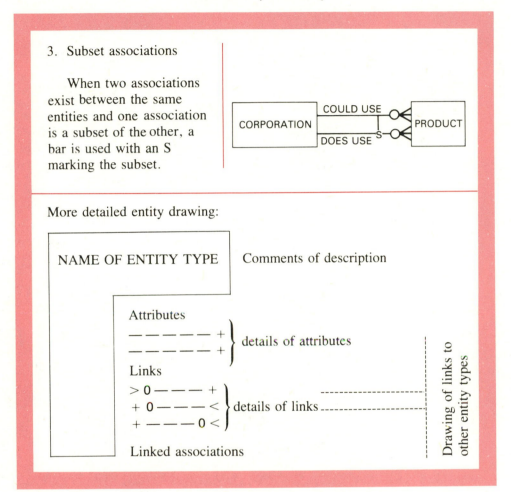

3. Subset associations

When two associations exist between the same entities and one association is a subset of the other, a bar is used with an S marking the subset.

More detailed entity drawing:

REFERENCES

1. James Martin, *Strategic Data-Planning Methodologies*, Prentice-Hall Inc., Englewood Cliffs, NJ, 1982.

2. From work done by Ken Winter, Rik Belew, and Bob Walter at Database Design Inc., Ann Arbor, MI.

3. From internal report by Ian Palmer, James Martin Associates, London.

21 DATA NAVIGATION DIAGRAMS

INTRODUCTION
Once a thorough, stable, fully normalized, data model exists, the task of the systems analyst and application designer is much easier [1]. The designer must determine how he *navigates* through the data base. He needs a clear diagramming technique for this. This chapter discusses the diagramming of data-base navigation.

A *data navigation diagram* is drawn on top of a data model. It can be converted into a program structure as represented in an action diagram. Data navigation diagrams are also sometimes called data access maps, or logical access maps when emphasizing that a logical data structure is used. They are sometimes drawn in a file environment as well as in a data-base environment.

DIVIDE AND CONQUER
A basic principle of structured design is *divide and conquer*. Complex, entangled designs need to be reduced to clean, relatively simple modules. The existence of a thoroughly normalized data model enables complex applications to be reduced to relatively simple projects which *enter and validate* data, *update* data, *perform computations* on data, *handle queries* and *generate reports, generate routine documents, conduct audits*, and so on. Sometimes these projects use data in complex ways with many cross-references among the data.

Most such projects can be performed *by one person*, especially when fourth-generation languages are used. The main communication among separate developers is *via the data model*. Most of the human communication problems of large programming projects can be made to disappear. Figure 21.1 illustrates this.

One-person projects are highly appealing. Management can select the person for the job and motivate him highly for speed and excellence. He is in charge of his own success. He is not a cog in a tangled human machine. He will not be

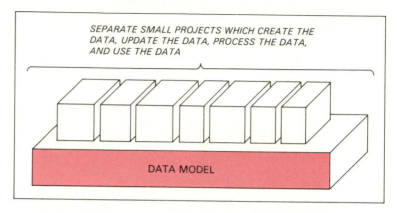

Figure 21.1   Well-designed, stable data model, with appropriated data management facilities, permits application development to be a series of separate quick-to-implement projects—mostly one-person projects. Communication among the projects is via the data model.

slowed down or have to rewrite his code because of other people. When he has finished, management can judge his results and reward him appropriately.

Although one person may be responsible for each of the blocks in Fig. 21.1, a team may sit side by side at terminals so that they can compare notes, see each other's displays, and help each other to understand the meaning of the data.

The divide-and-conquer strategy resulting in one-person projects is made stronger by the user of higher-level data-base languages, and fourth-generation languages in general. With those, one person can often obtain ten times the results in a given time, that he could with COBOL. A one-person team can replace a ten-person team. Most fourth-generation languages depend heavily on a data management facility. With COBOL or PL/1 also, subdividing projects into small modules is highly desirable and good data-base management assists this greatly.

SEPARATING DATA FROM PROCEDURES

The information-system needs of some big organizations have grown in complexity as data bases became established. It is possible to use the data in more complex ways. This increase in complexity can be handled only if there is an easy-to-use set of techniques for charting the way through the complexity.

A problem with many structured techniques is that they tangle up the structuring of the data with the structuring of the procedures. This complicates the techniques used. Worse, it results in data that are viewed narrowly and usually not put into a form suitable for other applications which employ the same data.

This and the following chapter assume that the data are separately designed using sound techniques, preferably automated. Data have properties of their own, independent of procedures that lead to stable structuring. The users or analysts who design procedures employ a data model and consider the program actions which use that data model. Often the data model is designed by a separate data administrator. Sometimes it is designed by the analysts in question.

The data navigation diagram is drawn on top of a portion of the data model and links it to the design of the programs which use those data. In this way it forms a simple, easy-to-use bridge between the data model and the procedure design. Any design of a data-base program should begin by sketching the navigation chart.

Some fourth-generation languages have enabled data to be used in more complex ways than previously. With some such languages, everything is oriented to the data-base structure. Many users, however, have difficulty learning how to use the full power of the language. They learn to formulate queries and generate reports but not how to handle complex data manipulation. One analyst described it as follows: "There is a threshold they cannot get through. It's like flying up through clouds and then all of a sudden the plane breaks out of the clouds and the sun shines." To get through this quickly needs appropriate, ultra-clear diagramming techniques, clearly taught.

DATA SUBMODEL

The first step in creating procedures that use a data model is to identify which entity types will be employed. The designer examines the overall entity-relationship diagram and indicates the entity types he expects to use. A subset diagram is created which shows only these entity types.

To help ensure that the designer has not forgotten any entity type which is needed, a *neighborhood* may be displayed. The neighborhood of one entity type is the set of entity types that can be reached from it by traversing one link in the entity-relationship diagram. Sometimes the data model may contain extra information saying that additional neighbors should be examined. The designer displays a list of such neighbors of the entity types he is interested in. Occasionally, the neighbor one link away may be a concatenated entity type (containing intersection data) and the designer inspects the other entity type(s) in the concatenated record.

The designer examines the neighborhood. He sees the entity types that his procedure will use, plus a few more. He eliminates those he does not want. There may be some which he would not have thought about if he had not displayed the neighborhood. There may be some which have *madatory* links to records he has specified. For example, when a *booking* record is created, the *seat inventory* record *must* be updated.

The designer then indicates for each entity-record type whether a CREATE, READ, UPDATE, or DELETE will occur. If a record is to be *created*, the designer must consider what relationship links must be built to other records.

These considerations give the designer a subset data model that his procedure will use. Usually, this is small enough to draw on one sheet of paper. Where possible it should be created by a computerized tool.

ACCESS SEQUENCE   The next step is to determine the sequence of accesses through this subset data mode. The designer may draw the sequence on the subset data model. He may indicate:

1. READ entity type A.
2. READ entity type B.
3. CREATE entity type C.

This might be drawn as follows:

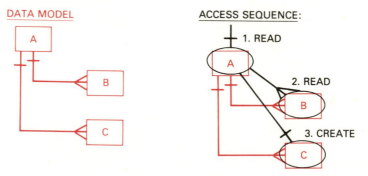

The link between the first READ and the second READ has a crow's-foot on it. This means that *many* entity type B records may be read in association with one entity type A record. Only one entity type C record is created and it is associated with the A record. This is shown by the one-to-one link to the creation of the C record.

The three accesses could be drawn on an action diagram as follows:

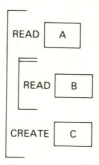

The data navigation diagrams need to be *automatically* convertible into the equivalent action diagram.

CONDITIONS

Often an access is made only if a certain condition applies. For example, in the processing of an order, the first step may be to check the customer's credit. If the credit is bad, the subsequent accesses are not carried out. This optionality is shown on the data navigation diagram using a broken line with a dot (as on other types of diagrams). The reader may read the dot as "o" for "optional."

In the diagram above, the third access, which creates record C, may be optional. We illustrate this as follows:

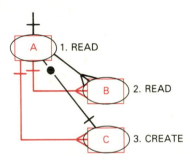

The dot, representing a condition, may be given a number, or a condition may be written against it, thus:

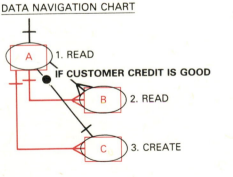

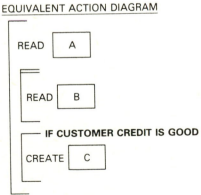

MUTUAL EXCLUSIVITY

As on other types of diagrams, two or more paths may be mutually exclusive. This is represented by a forking line with a dot at the fork:

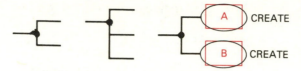

In the example above, the record created may be *either* entity type A *or* entity type B:

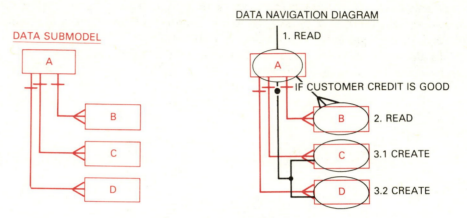

Mutually exclusive access paths translate into a case structure in an action diagram:

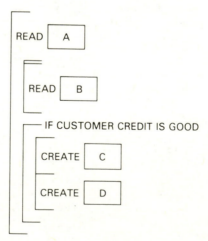

The conditions that control the case structure may be written on the navigation diagram, although often an action diagram is a more appropriate place to write this level of detail.

MUTUAL INCLUSIVITY

Again, two or more paths may be mutually *inclusive*; that is, if one is taken, the other must also be taken. This is represented by a forking line:

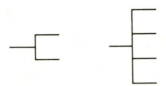

In the example below two records will be created if the customer's credit is good, one of entity type C and one of entity type D:

DATA NAVIGATION DIAGRAM

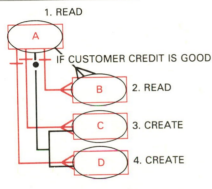

EQUIVALENT ACTION DIAGRAM

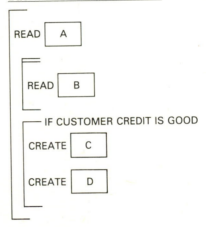

CONVERSION TO ACTION DIAGRAMS

The notation on the data navigation diagram includes the indicators for one-to-one and one-to-many (crow's-feet) associations, optionality (dots), mutual inclusivity, and mutual exclusivity that we have seen on other diagrams.

The navigation diagram should be convertible directly to an action diagram. If the navigation diagram is drawn using a computerized tool, this conversion should be *automatic*.

Figure 21.2 shows a variety of data navigation diagrams and their equivalent action diagrams. The designer adds information about calculations, printouts, and other procedures to the action diagram.

The process we describe in this chapter is ideal for a computerized graphics tool that enables the designer to extract a portion of a data model, explore its neighborhood, draw a navigation diagram on it, convert this to an action diagram,

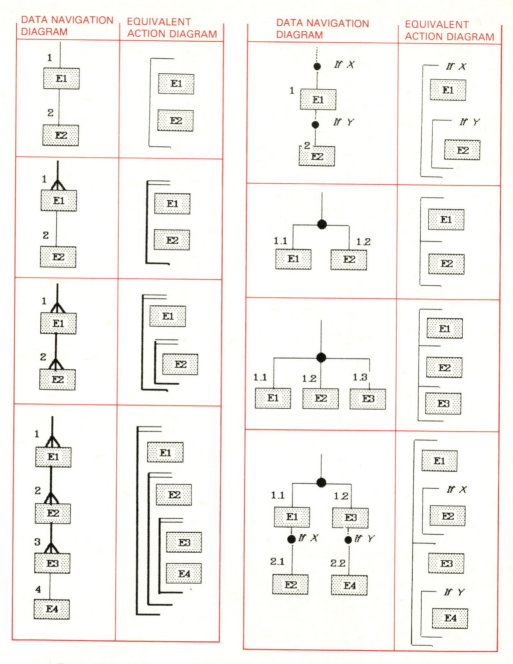

Figure 21.2 Variety of data navigation diagrams with their equivalent action diagrams.

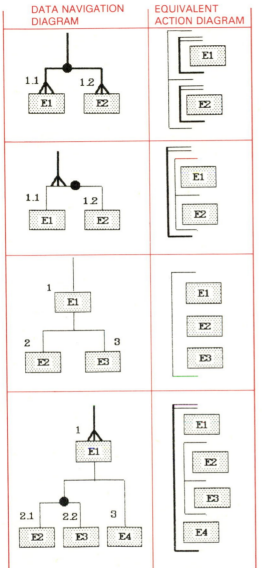

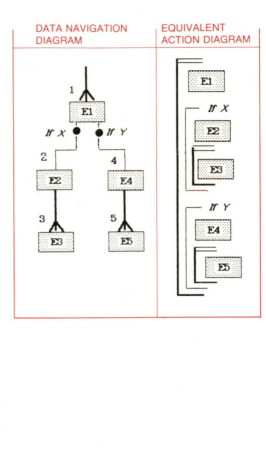

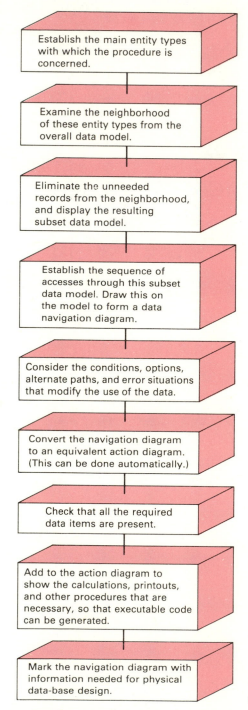

Establish the main entity types with which the procedure is concerned.

Examine the neighborhood of these entity types from the overall data model.

Eliminate the unneeded records from the neighborhood, and display the resulting subset data model.

Establish the sequence of accesses through this subset data model. Draw this on the model to form a data navigation diagram.

Consider the conditions, options, alternate paths, and error situations that modify the use of the data.

Convert the navigation diagram to an equivalent action diagram. (This can be done automatically.)

Check that all the required data items are present.

Add to the action diagram to show the calculations, printouts, and other procedures that are necessary, so that executable code can be generated.

Mark the navigation diagram with information needed for physical data-base design.

Figure 21.3 Steps for designing data-base applications.

add details to and edit the action diagram, and possibly generate executable code from the resulting action diagram.

Figure 21.3 lists the steps for designing data-base applications in this way. These steps are often applied manually. They also are ideal for the computerization of application building.

COMPUTERIZED DIAGRAMMING

The navigation diagrams so far in this chapter show hand-drawn rings around the entity types. For computerization, a less casual diagram is needed. Figure 21.4 shows a computer-drawn version of the diagram. It should be noted that the boxes and symbols on the action diagram are as similar as possible to those on the navigation diagrams.

DESIGN OF AN ORDER ACCEPTANCE SYSTEM

We will illustrate this technique with the design of an order acceptance application for a wholesale distributor. A third-normal-form data model exists, as shown in Fig. 21.5. The designer knows that the application requires CUSTOMER-ORDER records and PRODUCT records. The *neighborhood* of these includes the records

CUSTOMER-ORDER
CUSTOMER
ORDER-LINE
BACKORDER
INVOICE
PRODUCT
ORDER-RATE
QUOTATION
INVOICE-LINE-ITEM
PURCHASE-LINE-ITEM

The designer examines the data items in these records. The application does not need any data in the INVOICE, INVOICE-LINE-ITEM, PURCHASE-LINE-ITEM, or QUOTATION records. The ORDER-RATE record should be updated. The designer would have neglected the ORDER-RATE record if he had not printed the neighborhood. The designer decides, then, that he needs six records and creates a submodel containing these records as shown in Fig. 21.6.

DATA SUBMODEL:

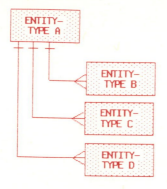

NAVIGATION DIAGRAM DRAWN BY HAND ON THE DATA MODEL:

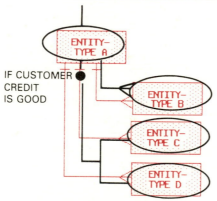

NAVIGATION DIAGRAM DRAWN BY
COMPUTER ON THE DATA MODEL:

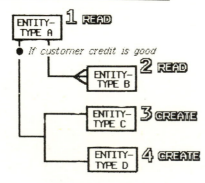

CORRESPONDING ACTION DIAGRAM

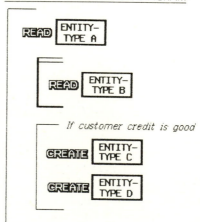

Figure 21.4 When the data navigation diagram is drawn on the data model using a computer graphics screen, the drawing is more formal than when drawn by hand.

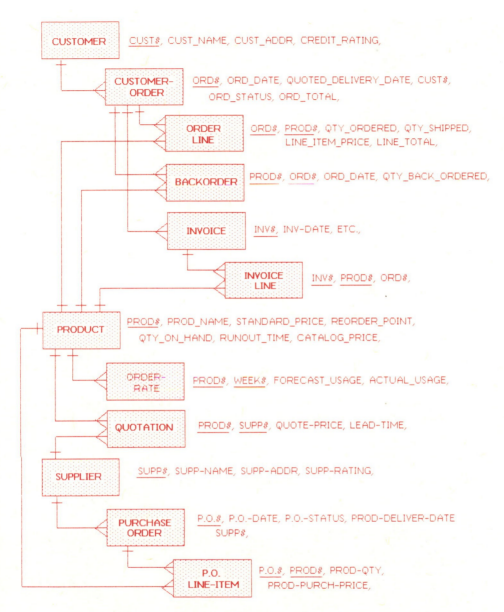

Figure 21.5 Data model for a wholesale distributor. Primary key data items are underlined. This model is not complete, but because it is correctly normalized it can be grown without pernicious impact to include such things as SALES-MAN, WAREHOUSES, ALTERNATE-ADDRESSES, and so on.

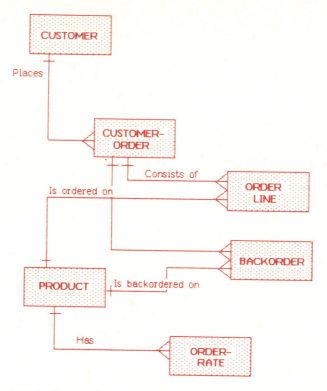

Figure 21.6   Subset of the data model in Fig. 21.5, extracted for the design of the order acceptance procedure.

Figure 21.7 shows his first drawing of the sequence in which the records will be accessed:

1. The CUSTOMER record will be inspected to see whether the credit-rating is good.

2. If the credit-rating is good, a CUSTOMER-ORDER record is created, linked to the CUSTOMER record.

3. For each product on the order, the PRODUCT record is retrieved to see whether stock of the product is available.

4. If stock is available, an ORDER-LINE record is created, linked to the CUSTOMER-ORDER record and PRODUCT record for that item.

5. The ORDER-RATE record for that PRODUCT is retrieved, a new order rate is calculated, and the ORDER-RATE record is updated.

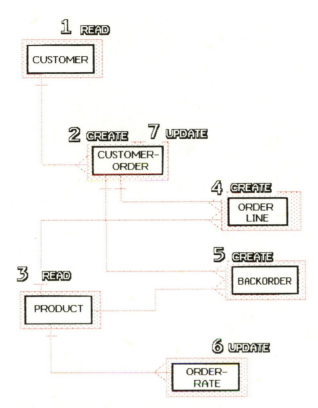

Figure 21.7   Preliminary indication of the sequence of accesses to the data, marked on top of the data model of Fig. 21.6.

6. If stock is not available, a BACKORDER record is created, linked to the CUSTOMER-ORDER and PRODUCT records.

7. When all items are processed, an order confirmation is printed and the CUSTOMER-ORDER record is updated with ORDER-STATUS, ORDER-TOTAL, and DELIVER-DATE.

   The first action is to read the CUSTOMER record and check the customer's credit rating. A condition dot is needed after the CUSTOMER record access. If the credit-rating is good, a CUSTOMER-ORDER record will be created *and* the PRODUCT records for the order will be read. There may be many PRODUCT records for the order, so a crow's-foot is used on the line showing the product access:

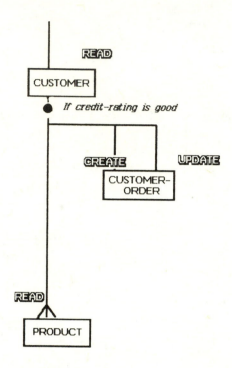

For each product there may or may not be stock available. This determines whether an ORDER-LINE record is created or a BACKORDER record. The designer, therefore, connects the PRODUCT access to the ORDER-ITEM and BACKORDER accesses with a mutually exclusive indicator:

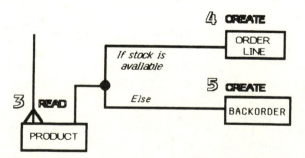

If an ORDER-ITEM record is created, the ORDER-RATE record for that product must be updated. A branching line goes to the ORDER-LINE access and to the ORDER-RATE access:

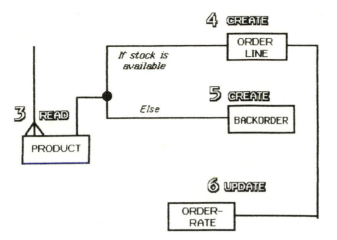

After the group of accesses above, the CUSTOMER-ORDER record is updated. Three sets of actions thus follow the checking of the customer's credit-rating:

1. Create the CUSTOMER-ORDER record.

2. Perform the actions above for each product on the order.

3. Update the CUSTOMER-ORDER record.

Figure 21.8 shows the resulting navigation diagram. This is represented as an action diagram in Fig. 21.9. The designer may now add more detail to the action diagram. Figure 21.10 shows the action diagram with the data items for each entity type displayed. The designer now represents the condition statements so that they use the data items:

IF CREDIT-RATING > 3

IF QTY-ON-HAND > 0

Figure 21.11 shows details of the calculations and printouts added to the action diagram.

FOURTH-GENERATION LANGUAGES   Figure 21.11 is independent of the programming language used. It is, however, designed to be as close as possible to the code representation used in fourth-generation languages and application generators. It can be converted di-

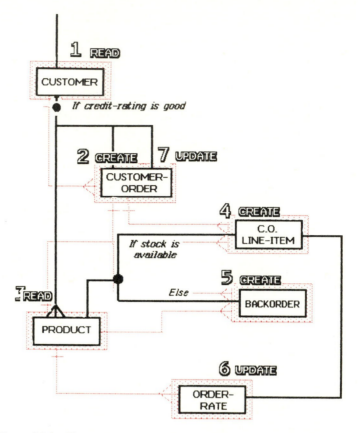

Figure 21.8   Figure 21.7 expanded into a full data navigation diagram.

rectly into languages such as FOCUS, RAMIS, MANTIS, NOMAD, IDEAL, CSP, and so on.

Figure 21.l2 shows its coding in APPLICATION FACTORY. Figure 21.13 shows its coding in IDEAL. It could also be coded in COBOL or PL/1 or it could form the basis for a tool that generates code in these languages. Fourth-generation languages or application generators of the future ought to assist an analyst or user in creating and manipulating diagrams, and then generate efficient executable code directly from the diagram.

FOUR STAGES   The procedure we have described has four main stages:

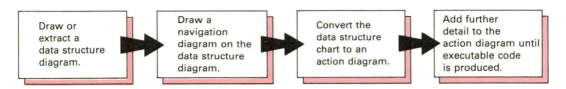

These stages are applied both to logical data models and to physical data structures. They are applied to logical data models when *specifying a procedure*, and to the more constrained data structures in a data-base management system when *designing a program*. For data-base management systems which directly represent the fully normalized data models, there is no difference between these.

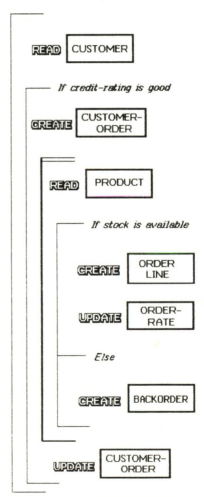

Figure 21.9 Action diagram (automatically) generated from the navigation diagram of Fig. 21.8.

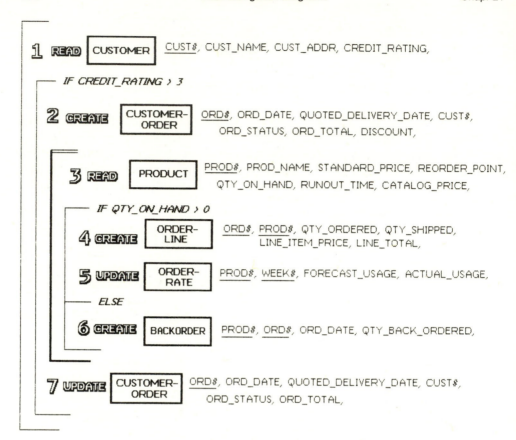

Figure 21.10   Action diagram of Fig. 21.9 annotated with the attributes of each entity type accessed. The condition statements are changed to appropriate attribute names.

Unfortunately, many data-base management systems employ physical data structures which deviate from the conceptual clarity of the entity-relationship model, and with these, the navigation diagram may be somewhat different from the navigation diagram drawn on the normalized data model.

LOGICAL AND PHYSICAL NAVIGATION DIAGRAMS

Navigation diagrams drawn on a logical data model are called *logical navigation diagrams*. Navigation diagrams drawn on a physical data structure are called *physical navigation diagrams*. Sometimes these are also called logical access maps and physical access maps.

The terms ''logical'' and ''physical'' ought to be clear when used about

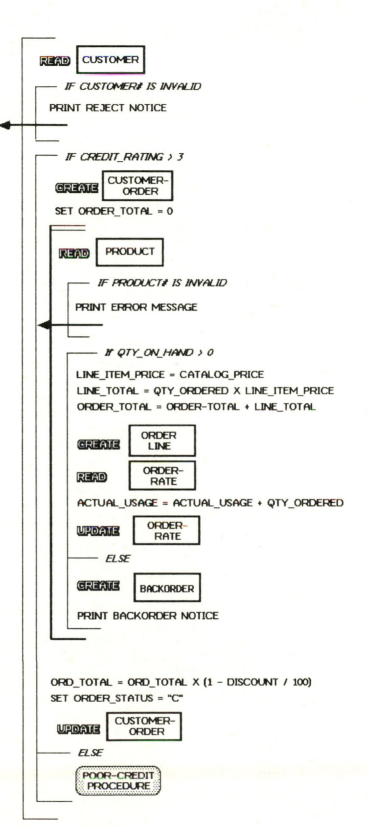

READ CUSTOMER
   IF CUSTOMER# IS INVALID
PRINT REJECT NOTICE

IF CREDIT_RATING > 3
CREATE CUSTOMER-ORDER
SET ORDER_TOTAL = 0

READ PRODUCT
   IF PRODUCT# IS INVALID
PRINT ERROR MESSAGE

   IF QTY_ON_HAND > 0
LINE_ITEM_PRICE = CATALOG_PRICE
LINE_TOTAL = QTY_ORDERED X LINE_ITEM_PRICE
ORDER_TOTAL = ORDER-TOTAL + LINE_TOTAL

CREATE ORDER LINE
READ ORDER-RATE
ACTUAL_USAGE = ACTUAL_USAGE + QTY_ORDERED
UPDATE ORDER-RATE

   ELSE
CREATE BACKORDER
PRINT BACKORDER NOTICE

ORD_TOTAL = ORD_TOTAL X (1 – DISCOUNT / 100)
SET ORDER_STATUS = "C"
UPDATE CUSTOMER-ORDER

   ELSE
POOR-CREDIT PROCEDURE

Figure 21.11 Action diagram of Fig. 21.10 expanded to show the calculations, printouts, and so on.

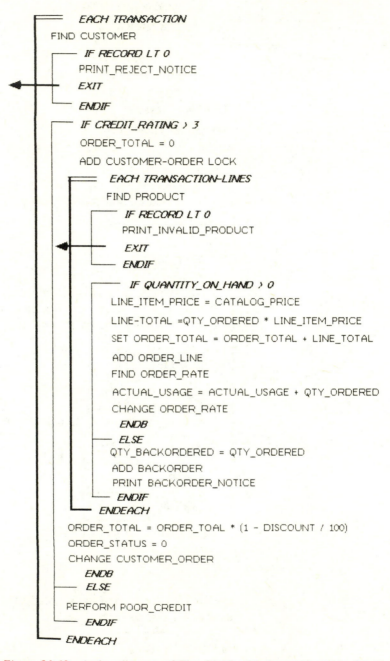

```
      EACH TRANSACTION
  FIND CUSTOMER
      IF RECORD LT 0
      PRINT_REJECT_NOTICE
      EXIT
      ENDIF
      IF CREDIT_RATING > 3
      ORDER_TOTAL = 0
      ADD CUSTOMER-ORDER LOCK
          EACH TRANSACTION-LINES
          FIND PRODUCT
              IF RECORD LT 0
              PRINT_INVALID_PRODUCT
              EXIT
              ENDIF
              IF QUANTITY_ON_HAND > 0
              LINE_ITEM_PRICE = CATALOG_PRICE
              LINE-TOTAL =QTY_ORDERED * LINE_ITEM_PRICE
              SET ORDER_TOTAL = ORDER_TOTAL + LINE_TOTAL
              ADD ORDER_LINE
              FIND ORDER_RATE
              ACTUAL_USAGE = ACTUAL_USAGE + QTY_ORDERED
              CHANGE ORDER_RATE
               ENDB
              ELSE
              QTY_BACKORDERED = QTY_ORDERED
              ADD BACKORDER
              PRINT BACKORDER_NOTICE
              ENDIF
          ENDEACH
      ORDER_TOTAL = ORDER_TOAL * (1 - DISCOUNT / 100)
      ORDER_STATUS = 0
      CHANGE CUSTOMER_ORDER
       ENDB
      ELSE
  PERFORM POOR_CREDIT
      ENDIF
  ENDEACH
```

Figure 21.12 Action diagram of Fig. 21.11 with the words of the fourth-generation language APPLICATION FACTORY. This is an executable program written in APPLICATION FACTORY.

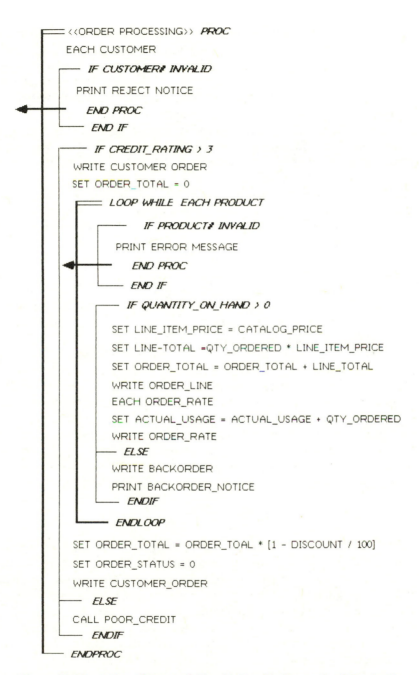

```
<<ORDER PROCESSING>> PROC
EACH CUSTOMER
    IF CUSTOMER# INVALID
    PRINT REJECT NOTICE
        END PROC
        END IF

        IF CREDIT_RATING > 3
    WRITE CUSTOMER ORDER
    SET ORDER_TOTAL = 0
            LOOP WHILE EACH PRODUCT
                IF PRODUCT# INVALID
                PRINT ERROR MESSAGE
                    END PROC
                    END IF

                IF QUANTITY_ON_HAND > 0
                SET LINE_ITEM_PRICE = CATALOG_PRICE
                SET LINE-TOTAL =QTY_ORDERED * LINE_ITEM_PRICE
                SET ORDER_TOTAL = ORDER_TOTAL + LINE_TOTAL
                WRITE ORDER_LINE
                EACH ORDER_RATE
                SET ACTUAL_USAGE = ACTUAL_USAGE + QTY_ORDERED
                WRITE ORDER_RATE
                ELSE
                WRITE BACKORDER
                PRINT BACKORDER_NOTICE
                    ENDIF
            ENDLOOP
    SET ORDER_TOTAL = ORDER_TOAL * [1 - DISCOUNT / 100]
    SET ORDER_STATUS = 0
    WRITE CUSTOMER_ORDER
    ELSE
    CALL POOR_CREDIT
        ENDIF
ENDPROC
```

Figure 21.13 Action diagram of Fig. 21.11 with the words of the fourth-generation language IDEAL. This is an executable program written in IDEAL.

data. "Logical" refers to data as perceived by the analyst or user; this representation of data should be designed for maximum conceptual clarity. "Physical" refers to data as stored in the machine; this representation of data is designed for machine performance. "Logical" is used to describe the fully normalized data model. "Physical" is used to describe the data structured for storage with pointers, chains, rings, tree structures, and so on. Unfortunately, however, there is much confusion about the word "logical" because vendors of data-base management systems such as IBM's IMS have misused the word "logical" to mean specific data structures which are not necessarily normalized, and which do not represent the clearly structured data model.

Nevertheless, it is important to understand that the physical data structures are often different from the fully normalized data model, so physical data navigation may be different from that which is conceived and drawn in the fully normalized data model.

PHYSICAL DESIGN

Initially, the physical aspects of accessing the data are ignored. The navigation diagram is drawn as though the data existed in memory for this application alone. Later the physical designer adjusts the navigation diagram as appropriate. To help the physical designer, the navigation diagram should be annotated with details of quantities of accesses.

The accesses on the physical navigation diagram may not be in the same sequence as the accesses on the logical navigation diagram. A logical diagram may have two updates of the same record type. Physically, both of these updates would be done at the same time. The first update may be done in computer main memory and not written on the external storage medium until the second update can be completed. Similarly, if a parent record has a child in a physical data base, the two may be updated together. A child record is not created physically until the parent is created, although it could appear first on a logical navigation diagram.

COMPUTERIZED HELP IN DESIGN

As the analyst goes through the steps of application design, there are certain questions that he should ask at each stage. This can be made into a formal procedure. If he is using a computerized tool for carrying out the design steps of Fig. 21.3, the machine should make him address the relevant design questions at each state. This leads to better quality design with well-thought-out controls, and should go as far as possible to automatically generating the next stage of the design until executable code is reached.

When extracting and editing the data submodel, the computer can show the designer the *neighborhood* of the entity types in question, and display their attributes. It can ask which types of entities are to be created, retrieved, updated, or deleted, as in Fig. 21.14. The designer replies and may receive a screen like

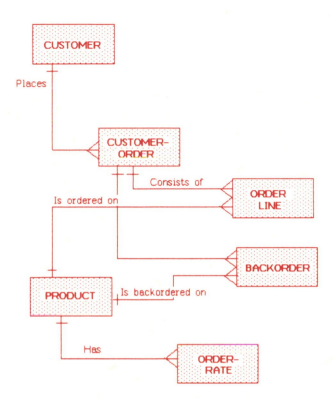

Figure 21.14   Figs. 21.14 to 21.21 show the building of a data navigation
diagram on an edited data submodel, using a personal computer screen. The
resulting navigation diagram can automatically be converted to an action diagram like that of Fig. 21.10.

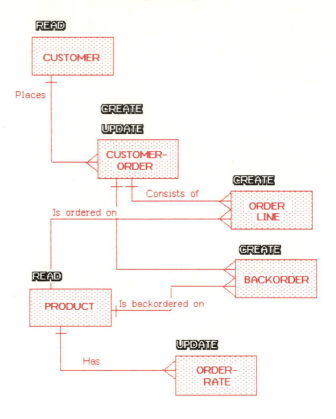

Figure 21.15

that in Fig. 21.15. He responds by indicating the sequence in which the accesses to data are carried out.

The computer asks the user questions about each access in turn. Figures 21.16 and 21.17 show it asking about the first access. It draws the first access shown in Fig. 21.17 and then asks what happens if the record in question cannot be found. All such error conditions need to be dealt with.

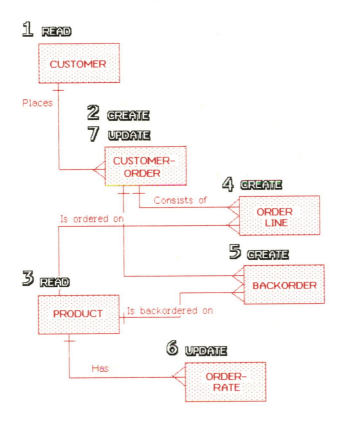

Figure 21.16

Figures 21.18 and 21.19 ask about the second access. The questions are somewhat different because this access *creates* a record. The designer responds in Fig. 21.18 by saying that the access occurs *SOMETIMES*. There Fig. 21.19 shows a *condition* dot and asks the designer to describe the condition. The designer enters "IF CREDIT-RATING IS GOOD."

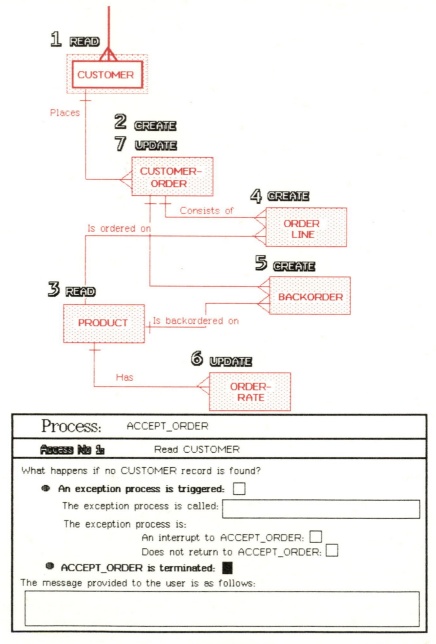

1 READ
CUSTOMER

Places

2 CREATE
7 UPDATE
CUSTOMER-ORDER

4 CREATE
Consists of
ORDER LINE

Is ordered on

5 CREATE
BACKORDER

3 READ
PRODUCT
Is backordered on

6 UPDATE
Has
ORDER-RATE

Process:	ACCEPT_ORDER
Access No 1:	Read CUSTOMER

What happens if no CUSTOMER record is found?

- **An exception process is triggered:** ☐

   The exception process is called: _____

   The exception process is:

            An interrupt to ACCEPT_ORDER: ☐
            Does not return to ACCEPT_ORDER: ☐

- **ACCEPT_ORDER is terminated:** ■

The message provided to the user is as follows:

_____

Figure 21.17

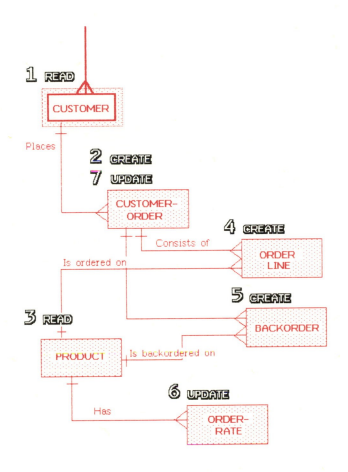

1 READ
CUSTOMER

Places

2 CREATE
7 UPDATE

CUSTOMER-
ORDER

Consists of

4 CREATE
ORDER
LINE

Is ordered on

5 CREATE
BACKORDER

3 READ
PRODUCT

Is backordered on

6 UPDATE
ORDER-
RATE

Has

Process: ACCEPT_ORDER		
Access No 2: Create CUSTOMER_ORDER		
For each CUSTOMER access the number of CUSTOMER_ORDER records created is: One: ■ Many: ☐		
CUSTOMER_ORDER records are created: Always: ☐ Sometimes: ■		
Access is: **Via CUSTOMER record:** ■ Other: ☐		

Figure 21.18

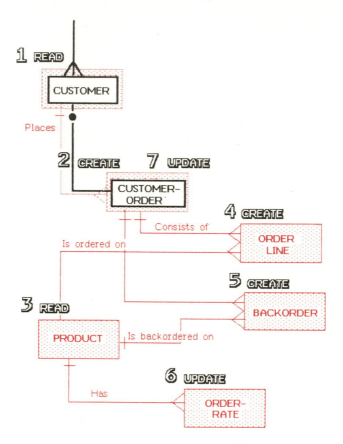

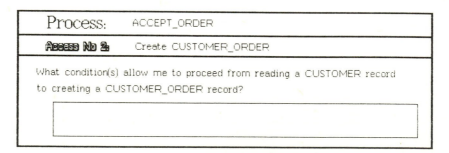

Figure 21.19

Figure 21.20 shows a later stage in the dialogue where the designer has described an UPDATE access. The panel shows the data items in the ORDER-RATE record and asks which will be updated. The designer indicates that AC-TUAL-USAGE will be updated, so the computer in Fig. 21.21 asks for the formula or process by which it is updated.

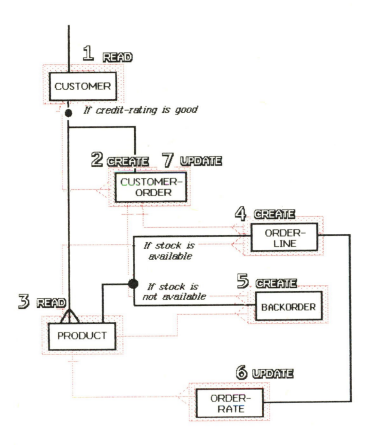

1 READ
CUSTOMER

If credit-rating is good

2 CREATE 7 UPDATE
CUSTOMER-
ORDER

4 CREATE
ORDER-
LINE

*If stock is
available*

*If stock is
not available*

5 CREATE
BACKORDER

3 READ
PRODUCT

6 UPDATE
ORDER-
RATE

Process: ACCEPT_ORDER

Access No 6:   Update ORDER_RATE

Which data-items will be changed in ORDER_RATE?

PRODUCT#   ☐
WEEK#   ☐
FORECAST_USAGE   ☐
ACTUAL_USAGE   ■

Are any other data-items needed in ORDER_RATE?   Yes: ☐   No: ■

Figure 21.20

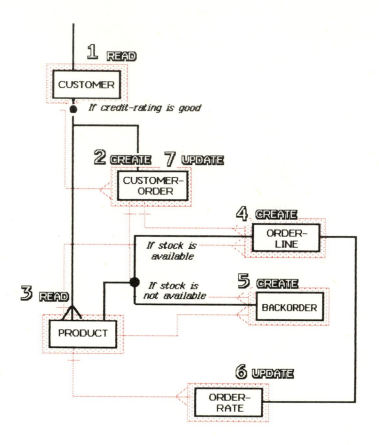

1 READ

CUSTOMER

If credit-rating is good

2 CREATE **7** UPDATE

CUSTOMER-
ORDER

4 CREATE

ORDER-
LINE

*If stock is
available*

3 READ

*If stock is
not available*

5 CREATE

BACKORDER

PRODUCT

6 UPDATE

ORDER-
RATE

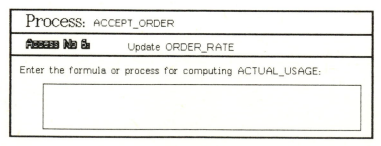

Process: ACCEPT_ORDER

Access No 6:   Update ORDER_RATE

Enter the formula or process for computing ACTUAL_USAGE:

Figure 21.21

The objective of such a design dialogue is to speed up the design and to ensure that all necessary controls and error conditions are thought about. The system should convert the results to an action diagram from which executable code can be created, as in Figs. 21.12 and 21.13. The dialogue should also collect figures which will assist the physical designer who is concerned with performance.

COMPLEXITY

The subset data model that is used for one application usually does not become very big. It can usually be drawn on one page or screen and so can be the associated navigation diagram.

In some corporations with highly complex data processing, the subset data models have never exceeded a dozen third-normal-form records. Most do not exceed eight.

STANDARD PROCEDURE

It is very simple to *teach* the use of data navigation diagrams. These should be an installation standard rather than a tool of certain individuals. Paper forms and guidelines are often used, rather than computer-aided design.

One giant aerospace corporation, where more than a thousand navigation diagrams were drawn, found that this approach highlighted the transaction-driven nature of good data-base usage. Transaction-driven design had been surprisingly difficult for many analysts to grasp because they had learned techniques which (like many structured techniques) were batch oriented.

REFERENCE

1. James Martin, *Managing the Data-Base Environment*, Prentice-Hall, Inc., Englewood Cliffs, NJ, 1983.

22 COMPOUND DATA ACCESSES

INTRODUCTION Traditional data-base navigation, as described in Chapter 21, uses *simple* data-base accesses: CREATE, READ, UPDATE, and DELETE. These carry out an operation on *one* instance of *one* record type.

Some high-level languages permit the use of statements that relate to not one but many instances of records and sometimes more than one record type. We will refer to these as *compound* data-base accesses. Examples of such statements are:

SEARCH

SORT

SELECT   (certain records from a relation or file)

JOIN   (two or more relations or files)

PROJECT (a relation or file to obtain a subset of it)

DUPLICATE

CREATE, READ, UPDATE, and DELETE may also be used to refer to multiple instances of a record type. DELETE, for example, could be used to delete a whole file.

Where a data-base access refers to one instance of record type, we have used a single box containing the name of the record type, thus:

Where a compound access is used which refers to more than one instance, we will use a double box containing the name of the record type or entity type.

Often this double box needs a qualifying statement associated with it to say how it is performed, for example:

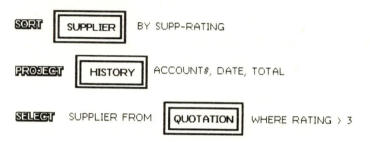

The type of operation is shown as previously.

RELATIONAL JOINS   A *relational join* merges two relations (logical files or tables) on the basis of a common field [1]. For example, the EMPLOYEE relation and the BRANCH relation might look like this:

BRANCH

BRANCH-ID	LOCATION	BRANCH-STATUS	SALES-YEAR-TO-DATE
007	Paris	17	4789
009	Carnforth	2	816
013	Rio	14	2927

EMPLOYEE

EMPLOYEE#	EMPLOYEE-NAME	SALARY	CODE	MANAGER	CITY
01425	Kleinrock	42000	SE	Epstein	Rio
08301	Ashley	48000	SE	Sauer	Paris
09981	Jenkins	45000	FE	Growler	Rio
12317	Bottle	91000	SE	Minski	Carnforth

These relations are combined in such a way that the CITY field of the EMPLOYEE relation becomes the same as the LOCATION field of the BRANCH relation.

We can express this with the statement

    BRANCH.LOCATION = EMPLOYEE.CITY

The result is, in effect, a combined record as follows:

PLOYEE#	EMPLOYEE-NAME	SALARY	CODE	MANAGER	CITY	BRANCH-ID	BRANCH-STATUS	SALES-YEAR-TO-DATE
425	Kleinrock	42000	SE	Epstein	Rio	013	14	2927
301	Ashley	48000	SE	Sauer	Paris	007	17	4789
981	Jenkins	45000	FE	Growler	Rio	013	14	2927
317	Bottle	91000	SE	Minski	Carnforth	009	2	816

This data-base system may not combine them in reality, but may join the appropriate data in response to queries or other operations. For example, if we ask for the MANAGER associated with each BRANCH-ID, the system will look up the BRANCH LOCATION for each BRANCH-ID, search for an EMPLOYEE.CITY data item with the same value, and find the MANAGER data item associated with that.

A JOIN is shown on a navigation chart by linking two or more entity-type access boxes together with a double line:

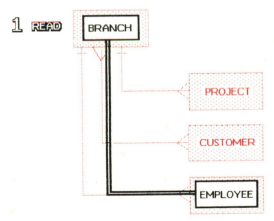

It is shown on an action diagram also by linking the boxes, with an access operation applying the combination:

A statement may be attached to the joined entity types showing how they are joined, like the BRANCH.LOCATION = EMPLOYEE.CITY above. Often this is not necessary because the joined entity records contain one common attribute which is the basis for the join. For example, the EMPLOYEE record probably contains the attribute BRANCH#, in which case we can simply show

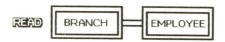

Using the *join* above we might say SELECT EMPLOYEE_NAME MANAGER, BRANCH_STATUS, CITY. The result is as follows:

EMPLOYEE-NAME	MANAGER	BRANCH-STATUS	CITY
Kleinrock	Epstein	14	Rio
Ashley	Sauer	17	Paris
Jenkins	Growler	14	Rio
Bottle	Minski	2	Carnforth

We might constrain the *join* operation by asking for employees whose code is SE and whose salary exceeds $40,000. The result would then be

EMPLOYEE-NAME	MANAGER	BRANCH-STATUS	CITY
Kleinrock	Epstein	14	Rio
Ashley	Sauer	17	Paris
Bottle	Minski	2	Carnforth

With the data-base language SQL from IBM, and others, this operation would be expressed as follows:

```
SELECT EMPLOYEE_NAME, MANAGER BRANCH_STATUS, CITY
FROM BRANCH EMPLOYEE
WHERE BRANCH.LOCATION = EMPLOYEE.CITY
AND CODE = SE
AND SALARY > 40000
```

This can be written on an action diagram as follows:

For a simple query such as this, we do not need a diagramming technique. The query language itself is clear enough. For a complex operation we certainly need to diagram the use of compound data-base actions. Even for queries, if they are complex, diagrams are needed.

AUTOMATIC NAVIGATION

A compound data-base action may require *automatic navigation* by the data-base management system. Relational data bases and a few nonrelational ones have this capability. For a data base without automatic navigation, a compiler of a fourth-generation language may generate the required sequence of data accesses.

With a compound data-base action, search parameters or conditions are often an integral part of the action itself. They are written inside a bracket containing the access box.

SIMPLE VERSUS COMPOUND DATA-BASE ACCESSES

There are many procedures that can be done with either simple data-base accesses or compound accesses. If a traditional DBMS is used, the programmer navigates his way through the data base with simple accesses. If the DBMS or language compiler has automatic navigation, higher-level statements using compound data-base accesses may be employed.

Suppose, for example, that we want to give a $1000 raise in salary to all employees who are engineers in Carnforth. With IBM's data-base language SQL we would write

```
UPDATE EMPLOYEE
GET SALARY = SALARY + 1000
WHERE JOB = 'ENGINEER'
AND OFFICE = 'CARNFORTH'
```

We can diagram this with a compound action as follows:

With simple actions (no automatic navigation) we can diagram the same procedure thus:

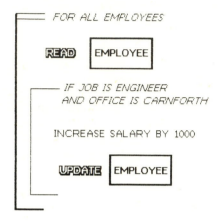

Similarly, a relational join can be represented with either a sequence of single actions or one compound action, as shown in Fig. 22.1. In this example, there are multiple projects of an EMPLOYEE-PROJECT record showing how employees were rated for their work on each project they were assigned to. They are given a salary raise if their average rating exceeds 6.

It ought to be an objective of nonprocedural languages to enable their user to achieve as much as possible without *separate* diagramming. The best way to achieve this may be to *incorporate the graphics technique into the language itself*. In other words, executable code is generated from the diagrams.

INTERMIXING SIMPLE AND COMPOUND ACTIONS

Sometimes compound and simple data-base actions are used in the same procedure. Figure 22.2 illustrates this. It uses the data structure shown in Fig. 21.5 and shows the process of reordering stock as it becomes depleted.

As a request for parts is satisfied, the quantity-on-hand, recorded in the PART record, is depleted. Each time this happens the program calculates whether to create an order for more parts from their supplier.

The requests for parts are handled interactively throughout the day. The person designing the procedure decides to create a temporary file of order requisitions, to accumulate such requisitions as they occur, and then to sort them and place the orders at the end of the day. In this way an order for many parts can be sent to a supplier, rather than creating a separate order each time a part reaches its reorder point.

Figure 22.2 shows the results. In the top part of Fig. 22.2, the selection of suppliers is done with a compound action. This can be represented with one

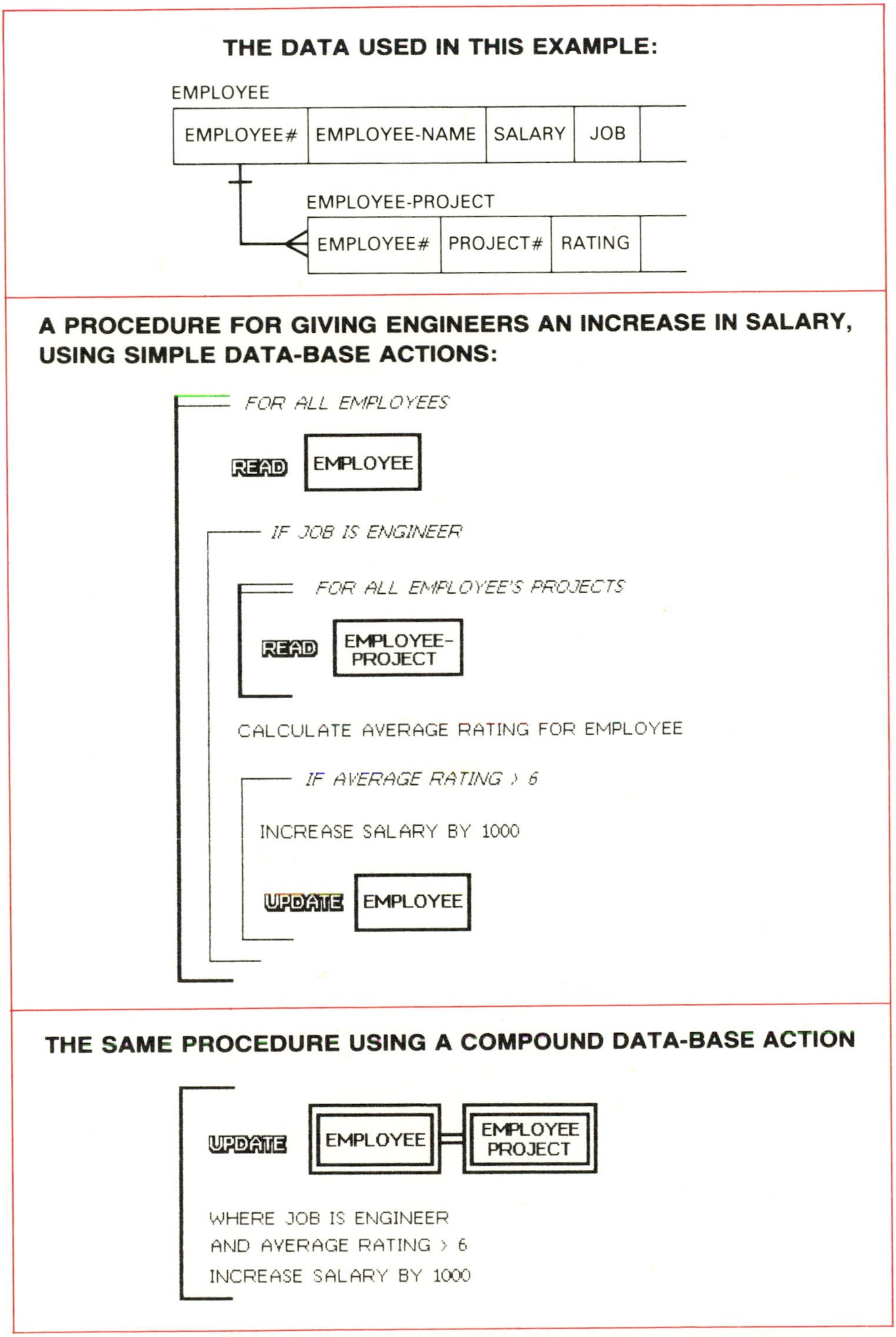

THE DATA USED IN THIS EXAMPLE:

EMPLOYEE

EMPLOYEE#	EMPLOYEE-NAME	SALARY	JOB	

EMPLOYEE-PROJECT

EMPLOYEE#	PROJECT#	RATING	

A PROCEDURE FOR GIVING ENGINEERS AN INCREASE IN SALARY, USING SIMPLE DATA-BASE ACTIONS:

FOR ALL EMPLOYEES

READ EMPLOYEE

  IF JOB IS ENGINEER

    FOR ALL EMPLOYEE'S PROJECTS

    READ EMPLOYEE-PROJECT

  CALCULATE AVERAGE RATING FOR EMPLOYEE

    IF AVERAGE RATING > 6

  INCREASE SALARY BY 1000

  UPDATE EMPLOYEE

THE SAME PROCEDURE USING A COMPOUND DATA-BASE ACTION

UPDATE EMPLOYEE — EMPLOYEE PROJECT

WHERE JOB IS ENGINEER
AND AVERAGE RATING > 6
INCREASE SALARY BY 1000

Figure 22.1   Illustration of a procedure that may be done with either multiple, simple data-base access commands or one compound access command action.

317

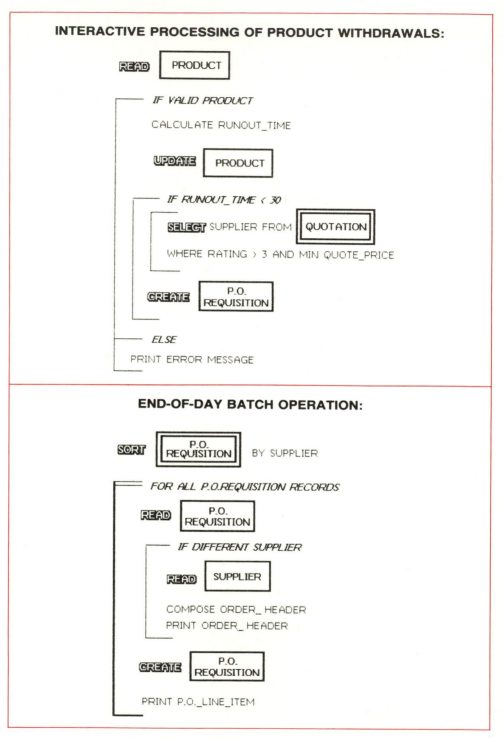

INTERACTIVE PROCESSING OF PRODUCT WITHDRAWALS:

READ PRODUCT

IF VALID PRODUCT

CALCULATE RUNOUT_TIME

UPDATE PRODUCT

IF RUNOUT_TIME < 30

SELECT SUPPLIER FROM QUOTATION

WHERE RATING > 3 AND MIN QUOTE_PRICE

CREATE P.O. REQUISITION

ELSE

PRINT ERROR MESSAGE

END-OF-DAY BATCH OPERATION:

SORT P.O. REQUISITION BY SUPPLIER

FOR ALL P.O.REQUISITION RECORDS

READ P.O. REQUISITION

IF DIFFERENT SUPPLIER

READ SUPPLIER

COMPOSE ORDER_HEADER
PRINT ORDER_HEADER

CREATE P.O. REQUISITION

PRINT P.O._LINE_ITEM

Figure 22.2 Procedures for product withdrawal and reordering using the data structure in Fig. 21.5 and a temporary P.O.-REQUISITION file. Two compound accesses are used in these procedures.

318

statement of some fourth-generation languages. For example, in SQL it might be

>SELECT SUPPLIER
>FROM QUOTATION
>WHERE RATING > 3
>AND MIN (QUOTE-PRICE)

In the bottom part of Fig. 22.2, the order requisition file is sorted by supplier. This can also be represented by one fourth-generation language statement:

>ORDER REQUISITION BY SUPPLIER

THREE-WAY JOINS   In some cases three-way joins are useful. Suppose that an accountant is concerned that accounts receivable are becoming too high. He wants to phone any branch office manager who has a six-month-old debt outstanding from a customer. The following record structures exist:

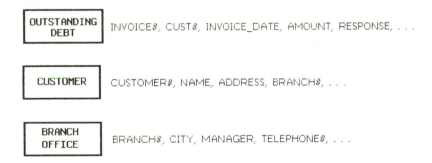

He enters the following query:

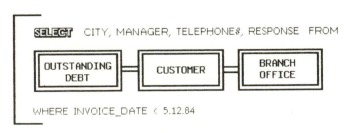

The three-way join is shown in a similar fashion to two-way joins. It could also be drawn on a data model to show a compound access in a navigation chart. Figure 22.3 shows this three-way join expressed with the language SQL.

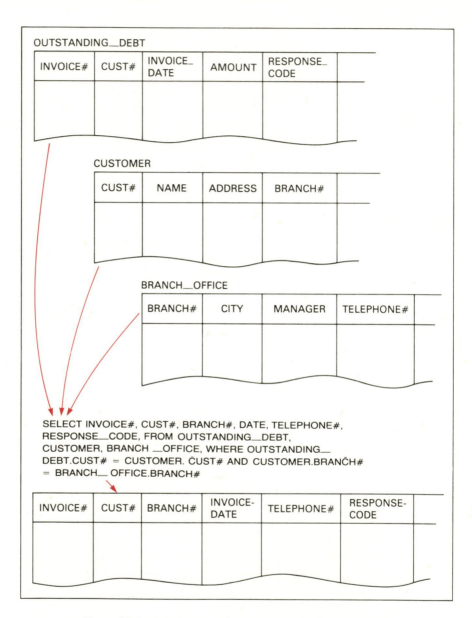

Figure 22.3   Join between three relations expressed with SQL.

SEMANTIC DISINTEGRITY

Unfortunately, compound accesses in high-level database languages sometimes give rise to subtle problems.

A user may enter a query with an easy-to-use query language; the query looks correct and the results look correct, but the results are in fact wrong.

The query with a triple join above is correct because OUTSTANDING__ DEBT is associated with *one* CUSTOMER and CUSTOMER is associated with *one* BRANCH. In the data model, these associations are as follows:

Suppose, however, that one CUSTOMER can be served by more than one BRANCH__OFFICE:

Then the use of the join is incorrect; there is semantic disintegrity in the query. The accountant might be phoning a branch manager who is not responsible for a customer's debt.

Again, suppose that two relations were joined as follows:

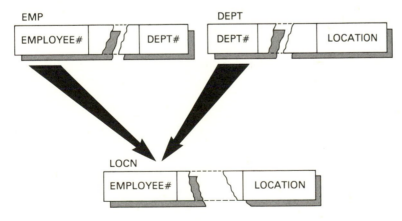

This join is valid if there is a one-to-one association between DEPT# and LO-CATION. It is not valid if there is a one-to-many association between DEPT#

and LOCATION, because although a department can have more than one location, an employee works in only one location. Drawing data items as ellipses, we have the following associations between data items:

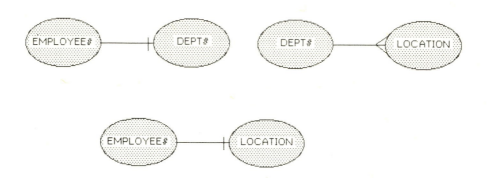

NAVIGATION PATHS

We can understand what a query or relational operation is doing by drawing a navigation path. To perform the join above, we start with EMPLOYEE# and find the associated DEPT#. For that DEPT# we find the associated LOCATION. We can draw this navigation path as follows:

Here we have only one-to-one paths so that there is no problem.

If, however, there were a one-to-many path from DEPT# to LOCATION, we would draw

This is invalid because there is *one* LOCATION, not many, for one employee.

We have the possibility of semantic disintegrity if the navigation path has a one-to-many link which is not the first link. For example,

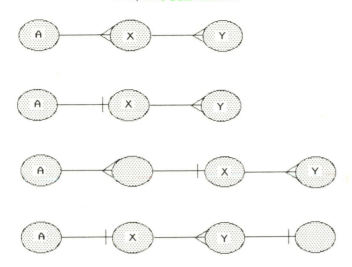

Many values of Y are associated with X, but they might not all be associated with A.

We do not necessarily know whether such a navigation path will be valid or not. Figure 22.4 shows two queries employing a JOIN. Their data and navigation paths are similar in structure. Both use fully normalized data. The one-to-many path makes the bottom one invalid, but not the top one. Because the software cannot tell for sure, it should warn the user that the results might be invalid.

To ensure integrity in relational operations, it is essential that the data are correctly and completely modeled. The designer can understand the effect of compound navigation by drawing appropriate diagrams. Diagrams showing details of the navigation path can warn of the danger of semantic disintegrity. The author has discussed semantic disintegrity more fully elsewhere [2].

FOURTH-GENERATION LANGUAGES

Different fourth-generation languages or high-level data-base languages have different dialects [3]. It would be useful if vendors of such languages would draw illustrations of the set of control structures and compound data-base accesses which their languages employ.

A compound data-base access is rather like a macroinstruction which is decomposed into primitive instructions by a compiler or interpreter before it is executed. To clarify how a compound data-base access in a language operates, the vendor might draw a diagram decomposing it into simple accesses. This is not always useful because some compound accesses are easy to understand but difficult to draw in a decomposed form (e.g., SORT).

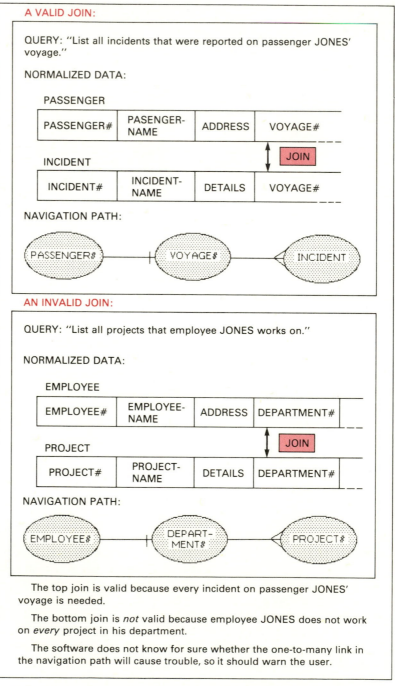

A VALID JOIN:

QUERY: "List all incidents that were reported on passenger JONES' voyage."

NORMALIZED DATA:

PASSENGER

PASSENGER#	PASENGER-NAME	ADDRESS	VOYAGE#

INCIDENT

INCIDENT#	INCIDENT-NAME	DETAILS	VOYAGE#

JOIN

NAVIGATION PATH:

PASSENGER# — VOYAGE# — INCIDENT

AN INVALID JOIN:

QUERY: "List all projects that employee JONES works on."

NORMALIZED DATA:

EMPLOYEE

EMPLOYEE#	EMPLOYEE-NAME	ADDRESS	DEPARTMENT#	

PROJECT

PROJECT#	PROJECT-NAME	DETAILS	DEPARTMENT#	

JOIN

NAVIGATION PATH:

EMPLOYEE# — DEPART-MENT# — PROJECT#

The top join is valid because every incident on passenger JONES' voyage is needed.

The bottom join is *not* valid because employee JONES does not work on *every* project in his department.

The software does not know for sure whether the one-to-many link in the navigation path will cause trouble, so it should warn the user.

Figure 22.4 Two queries using a join, which have similar data structures.

Compound accesses, like simple accesses, need to be converted directly into the code of very high level languages. Often the wording inside the bracket should resemble the resulting code. The dialect of the language is thus incorporated into the diagram. Converting the action diagrams to code can be computer assisted and ought to form part of an interactive design tool.

The software designer creating a fourth-generation language would do well to start with navigation diagrams and action diagrams, design an easy-to-use technique for charting procedures, and then create an interpreter or compiler with which code can be generated from the diagrams. The system analyst needs a computer screen which enables him to rapidly edit diagrams and add more detail until working code is created.

REFERENCES

1. James Martin, *Managing the Data-Base Environment*, Prentice-Hall, Inc., Englewood Cliffs, NJ, 1983.

2. James Martin, *Program Design Which Is Provably Correct*, Savant Technical Report 28, Savant Institute, Carnforth, Lancashire, UK, 1982.

3. James Martin, *Fourth-Generation Languages*, Savant Institute, Carnforth, Lancashire, UK, 1983.

23 A CONSUMER'S GUIDE TO DIAGRAMMING TECHNIQUES

INTRODUCTION

As we have seen, there are various competing diagramming techniques, many of them associated with structured methodologies. This chapter gives our views on those which we recommend. It repeats and summarizes some of the comments made in earlier chapters.

A mature systems analyst ought to be familiar with all of the diagramming techniques we have discussed, and immediately recognize their equivalences.

Figure 23.1 shows what each diagramming technique can draw. No technique can draw everything that is needed, so an installation needs a carefully selected combination of techniques.

Figure 2.1 showed an eight-area diagram representing the territory for which the diagramming is needed. At the top were the high-level overviews; at the bottom, detailed program structures. At the left were data; at the right, activities which used those data. Figure 2.2 showed where the techniques described in this book fit on that diagram.

In any one corporation, a choice is needed of techniques shown in Fig. 2.2. A corporation needs to standardize on an appropriate group of affiliated techniques.

CRITERIA FOR CHOICE

The structured revolution first gained momentum in the mid-1970s, and that is when many of the diagramming techniques evolved. A decade later the requirements of diagramming techniques had changed in certain important ways. Just as there was a change from flowcharts (1960s) to structure charts (1970s), so there needs to be a change from third-generation (1970s) to fourth-generation (1980s) techniques. Just as many old flowcharters refused to change or even accept the need for change, so many practitioners of the 1970s techniques also

WHAT CAN BE DRAWN WITH THE TECHNIQUE?	DECOMPOSITION DIAGRAMS	DEPENDENCY DIAGRAMS	DATA FLOW DIAGRAMS	ENTITY-RELATIONSHIP DIAGRAMS	DATA STRUCTURE DIAGRAMS	DATA NAVIGATION DIAGRAMS	HIPO CHARTS	STRUCTURE CHARTS	WARNIER-ORR CHARTS	MICHAEL JACKSON CHARTS	FLOWCHARTS	STRUCTURED ENGLISH	NASSI-SHNEIDERMAN CHARTS	ACTION DIAGRAMS	DATA-BASE ACTION DIAGRAMS	DECISION TREES AND TABLES	STATE-TRANSITION DIAGRAMS	HOS CHARTS
PROCESS STRUCTURES																		
Enterprise model showing corporate functions	YES							YES	YES					YES	YES			
Functional decomposition species I (tree structure only)	YES						YES	YES	YES	YES				YES	YES			YES
Functional decomposition species II (tree structure plus input and output)														YES	YES			YES
Functional decomposition species III (axiomatic control of decomposition)																		YES
Interaction between business events		YES	YES											YES	YES			YES
Flow of data		YES	YES											YES	YES			YES
Nonprocedural (compound) data-base actions															YES			
Control structures — Sequence	YES	YES						YES	YES		YES	YES	YES	YES	YES			YES
Control structures — Conditions	YES	YES									YES	YES	YES	YES	YES	YES	YES	YES
Control structures — Case structure		YES									YES	YES	YES	YES	YES	YES	YES	YES
Control structures — Repetition	YES	YES						YES	YES		YES	YES	YES	YES	YES			YES
Control structures — Loop control											YES	YES	YES	YES	YES			
Good for showing complex logic												YES	YES	YES		YES	YES	
Designed for showing highly complex decisions																YES	YES	
Linkage to data model															YES			YES
Linkage to fourth-generation languages														YES	YES			
DATA STRUCTURES																		
Tree structured data				YES	YES				YES	YES								YES
Plex structured data				YES	YES													YES
Derived data items					YES													
Corporate data models				YES	YES													
Data-base navigation						YES												

Figure 23.1   Summary of the capabilities of diagramming techniques.

have dug in their heels. Unfortunately, most systems analysts and programmers are still being trained to use older techniques with less computerized assistance, less end-user involvement, less verification before programming, and lower-productivity languages.

The following are important characteristics of today's diagramming methods:

User Friendly

There needs to be increasing end-user involvement in computing. Users should be able to read, check, and critique systems analysts' plans. Increasingly, they should be able to create applications themselves with user-friendly languages, or sketch applications which they require. User-friendly diagramming techniques are a major aid to spreading computer literacy.

Designed for Computerized Tools

Computer-aided systems analysis and computer-aided programming (CASA/CAP) speed up the building of systems, improve the quality of results, and make systems easier to test and to change (maintain). The diagramming technique should be designed for ease of sketching and editing at a computer screen. The computer should fill in many details by using a data dictionary, a data model, drawing techniques, verification techniques, and fourth-generation-language constructs.

Code Generation

It is desirable that wherever possible program code should be generated automatically. Some diagramming techniques do not help with this. Some are used to generate code skeletons showing the control structure of programs. Some can generate executable code.

One Tool Top to Bottom

Some diagramming techniques are appropriate for a high-level overview of systems or programs. Some are appropriate for showing low-level detail with conditions, case structures, and loop structures. It is desirable to have one program representation which can be decomposed steadily from the highest overview to the finest detail. This representation should encourage design refinement a step at a time with computerized editing. Changing representation techniques as the design is decomposed into detail encourages the making of errors.

Rigorous

Any possible method should be used to detect and eliminate errors as the system is being specified, designed, and programmed. Some diagramming techniques make possible a high level of automated checking; some do not. This is especially important in the creation of highly complex specifications. Some design techniques, though structured, have resulted in specifications which are full of errors.

Data-Base Oriented

Data-base technology is of great importance and is the foundation stone of most future data processing. Some structured techniques in common use represent files well, but not the nonhierarchical structures found in data bases. Some do not represent interactive data-base processing well. Good representation of data-base interactions and the use of data models is vital.

Oriented to Fourth-Generation Languages

Fourth-generation languages give much faster application development and avoid some of the structuring problems inherent in earlier language. Diagrams of structured designs should be decomposable into fourth-generation language code, with automated help if possible.

Speed

Productivity of application creation becomes more important as computers outnumber programmers by an ever-increasing ratio. Diagramming techniques are needed which are fast to use, both on paper and with computerized tools.

DATA AND PROCESSES

It is important to have clear diagrams of the structure of data. The design of processes and program structures is easier and more satisfactory if the data design already exists. Some structured techniques have used the same method for drawing data as for drawing processes. This is done in Jackson methodology and Warnier–Orr methodology. It works only for hierarchical data. Hierarchical data structures and hierarchical program structures can be drawn with the same technique. However, many data are network-structured by their very nature, not hierarchically structured. A different technique is needed to represent such data.

It is sometimes argued that the data extracted from a data base and used by a programmer are hierarchical; that is, the programmer's logical view of data is hierarchical. However, to impose this constraint tends to prevent the design team from understanding and using data-base navigation and relational concepts.

The data-base world needs correct data modeling. Data models are inevitably drawn differently from program structures. The analyst and programmer need to be thoroughly comfortable with the technique for representing systems and programs as well as that for representing data models. They need to be able to link the data models to their programs. Without this combination, any structured methodology is inadequate today.

DATA FLOW DIAGRAMS

Diagrams that show the passage of data or documents among separate processes are valuable. They enable analysts to draw and understand how complex operations use data. They can draw the flow of work tickets, requisitions, and so on, in a factory, or the flow of transactions in a banking system. At a lower level they can chart the flow of data among computer programs and modules.

Data flow diagrams are easy for both analysts and end users to understand. They have an important role to play. Data flow diagrams need linking to both the data model that is used and the program structures that result from the diagram. A concern with data flow diagrams is that they are deceptively simple. They can look correct and give the analyst a comfortable feeling, when a closer examination of the detail would reveal a different picture. Many specifications for complex systems have been created by analysts drawing large numbers of data flow diagrams leveled (nested) into many layers. When these specifications have been reexamined with more rigorous methods, they have been revealed to be full of inconsistencies, omissions, and ambiguities.

The appropriate role of the data flow diagram needs to be recognized. It is a useful form of overview sketch. It needs to be used in conjunction with thorough data analysis and data modeling.

A data flow diagram as commonly drawn cannot be converted automatically into a program structure. (In spite of misleading claims, this is a manual conversion needing human additions.) A more useful sequence is to draw a dependency diagram (which looks somewhat similar to a data flow diagram) and convert this automatically to an action diagram with appropriate software.

The flows of data marked on a data flow diagram often do not actually flow as indicated. Often, one process updates a data base and another process uses those data.

As commonly drawn, data flow diagrams are not rigorous; they are a sketch that is a useful aid to understanding. They need to be linked to rigorous techniques for data modeling and creation of program structures. For complex specifications this *must* be the species II or species III decomposition discussed in Chapter 8, and the checking associated with these forms of decomposition needs to be computerized. The human brain inevitably makes mistakes in specifications that are not computerized, and with complex specifications the quantity of mistakes is horrifying.

COMPREHENSIVE CAPABILITIES

For the design of processes or programs it is desirable to select one diagramming technique which can do as much as possible. If it can diagram some aspects of programs and not others, it is going to cause an unnatural break in the design sequence. Worse, it may constrain the thought processes of the designer.

Of the techniques in Figure 23.1, action diagrams and HOS charts are the most comprehensive. Each of these was specifically designed to give top-to-bottom design which could progress from the highest-level sketch of a system down to a level of detail from which code can be generated.

The other techniques have to switch horses in midstream. The most common methodology in North America is the combination of structure charts and pseudocode. In Europe, Michael Jackson diagrams and his ''structured text'' form of pseudocode are more common in Europe than in America. Another combination

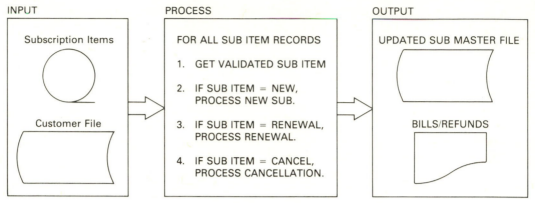

Overview HIPO diagram for the DISPATCH component of the subscription system.

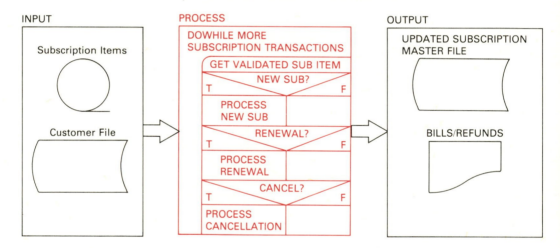

Figure 23.2   A HIPO diagram with a Nassi–Shneiderman diagram replacing its central process description.

is Warnier–Orr diagrams and pseudocode. Nassi–Shneiderman diagrams can replace pseudocode, but not a higher-level decomposition. Figure 23.2 shows a HIPO chart with a Nassi–Shneiderman diagram replacing its central process description.

Action diagrams avoid these clumsy combinations of basically different techniques. As shown in Figs. 16.5 to 16.7, one technique can show the highest-level functional decomposition and can be successively decomposed until executable code is reached. The labeling on the action diagram can be adapted to the dialect of a fourth-generation language. As shown in Chapters 21 and 22, the action diagram can be specifically related to data-base actions, including the compound actions of nonprocedural languages.

With software on a personal computer or terminal the designer can draw, edit, and change the action diagram quickly. He can link it to a data dictionary, data model, library of previously designed routines, screens, data structures, and so on. He may have the interpreter of a fourth-generation language (or third-generation, such as BASIC, C, or Pascal) running at the same personal computer or terminal.

ULTIMATE DECOMPOSITION   To be fully comprehensive the diagrams of processes should be extendable all the way from a high-level overview down to executable code in a fourth-generation language as in Fig. 16.7.

SPEED OF DRAWING   Productivity of analysts and programmers is becoming increasingly important as computers proliferate and the people who can create software for them remain scarce. We need techniques which are as fast to use as possible.

Diagramming techniques which are excessively time consuming can be frustrating to programmers under pressure who are attempting to achieve results quickly. Any diagram of complex procedures is likely to be redrawn many times. It should be quick to redraw. The importance of this was felt by us particularly when redrawing Nassi–Shneiderman diagrams. This is a time-consuming technique. Complex, hand-drawn Nassi–Shneiderman diagrams take a long time to change. Warnier–Orr diagrams and action diagrams, on the other hand, are quick and easy to change by hand. Decision tables can represent a complex family of conditions quickly and thoroughly. The same conditional logic would be time consuming to draw with flowcharts or Nassi–Shneiderman charts.

Diagramming, or more important, successive modification of diagrams, can be speeded up with an appropriate computerized tool. A diagramming method is needed which is good for easy computerized creation and modification of diagrams on inexpensive personal computers. The best computer-aided-design software can draw almost any diagram but it is expensive. A structured design technique should itself be designed so that it is not expensive to create good software for it.

Whether or not computerized tools exist, analysts, programmers, and increasingly end users argue with sketches on scrap paper and blackboards. A technique is needed which is good for sketching and which can progress to computerized representation. When computer graphics are used, the sketch can be automatically extended and dressed up to make it more respectable and clear to read. We are certainly going to have elegant computer graphics, but it would be a mistake to assume that this will completely replace hand drawing. The diagramming techniques used need to be appropriate for hand sketching as well as computer-aided design.

In spite of computer screens it is necessary to print diagrams that designers and programmers can take home, mail, scribble notes on, and so on. We stressed in Chapter 3 the need to obtain printouts on normal-sized paper rather than wide charts like architects' drawings. The diagrams should be printable on the inexpensive printer of a personal computer. This requires charts that spread out vertically rather than horizontally. It requires techniques that avoid entangled birds' nests of diagrams.

Many of the structure charts, data flow diagrams, and entity-relationship diagrams that we examined did not meet these criteria. They *could* meet them with different drawing techniques, as shown in Chapter 3. Modularization and multilevel nesting of diagrams can make them printable on small paper but tend to give excessive fragmentation which make them difficult to follow. Action diagrams were designed to be printable with inexpensive desktop printers.

Some analysts, programmers, and their managers love gigantic wall charts with which they can impress their associates. The desire to impress sometimes overrides the more practical concerns of obtaining results in a quick-and-easy fashion.

INTEGRITY CHECKING

The computer is a wonderful tool for integrity checking. It can with infinite patience correlate input and output, check data against a dictionary, and apply axioms for verification of correctness. Diagramming methods, and design techniques in general, are needed which facilitate computerized verification to the maximum extent. The larger and more complex the specification, the greater the need for computerized verification. Where computerized verification has been applied late in the history of complex specifications it has revealed an appalling mess. The human brain simply does not achieve the tedious correlation of detail that is needed, and a large team of human brains with human communication problems is much worse. Errors and inconsistencies tend to multiply as the square of the number of people in the team.

Most structured design uses hierarchical decomposition. We stressed in Chapter 8 that there are three species of hierarchical decomposition. Species I has no rules and is by far the most common. The computer can do little to help check its accuracy. Species II correlates and checks input and output data types throughout the entire design. With this, computerized help is of great value for complex systems. Species III allows only certain types of decomposition which are rigorously checked so that the entire resulting structure is internally correct.

Species II decomposition is easy to apply. It seems surprising that this level of automated checking is not widespread. Species III is more difficult to apply (although less difficult than the task of programming!). It requires a new level of rigor in thinking about systems, which appears to be a cultural shock to traditional analysts and programmers. Its payoff, however, is great because it permits a small team, often one person, to create complex specifications, from which correct code is automatically generated. The last column of Fig. 23.1 relates to HOS charts [1].

CODE GENERATION

Computer-aided diagramming and validation are highly desirable, but the computer can help in a more powerful way. Wherever possible it should generate code automatically.

Various methodology vendors claim to have automatic code generation. It is important to distinguish between two types of code generation:

1. Generation of a program skeleton giving the structure of the program but not the detail

2. Generation of executable code

Most methodologies do the former and not the latter. Clearly, the latter is much more valuable. The reader should be cautioned that many claims of code generation mean the former, not the latter.

Executable code is generated from HOS charts. Tools exist for generating logic modules from decision tables. Action diagrams can be linked to fourth-generation languages for creating executable code.

USER FRIENDLINESS

To achieve increased end-user involvement in computing, the types of diagrams used need to be easy to read and understand. As we discussed in Chapter 3, they should be as self-explanatory as possible. In many cases, end users should be taught to think about systems with the diagramming technique as a basic aid to clear thinking.

End users learn to read and draw some types of diagrams easily, but not others. Users can quickly understand data flow diagrams. Warnier–Orr diagrams are user-friendly except for their use of symbols such as " + ", "⊕", and (1, N). These symbols are quickly learned but to untrained users they appear to be difficult hieroglyphics. Action diagrams such as Figs. 16.2 to 16.4 are very easy to read by end users and can be expanded into successively more detail, encouraging the end user to understand the process of creating detailed applications.

Diagrams that permit the most complete verification are often the least user friendly. Of the techniques in Fig. 23.1 the most difficult for users to understand are Michael Jackson methodology and HOS methodology. With computerized tools for diagramming and verification it is possible to build methodologies which are both user friendly and rigorous. It is highly desirable that computer methodologies progress in this way.

Systems analysts and business analysts will increasingly use computer graphics to create, edit, change, catalog, and maintain the diagrams they draw. It is important to distinguish between two types of computer graphics:

1. Manual-Substitute Graphics

This type of graphics tool merely automates the drawing of diagrams that could be drawn by hand. It speeds up a manual methodology. Examples are MCAUTO's STRADIS/DRAW, and InTech's EXCELERATOR.

2. Power-Tool Graphics

This type of graphics performs extensive computations relating to the meaning of what is drawn. It may perform cross-checks, verifications, or design calculations. It may translate sketches into more complete designs asking the designer to make decisions or provide information. Examples are HOS, data-base design tools, and tools for creating specifications from which code can be generated.

The first type of graphics is anchored to design techniques which are commonly performed manually. The second makes possible design techniques which would be far too complex or difficult for most analysts to perform by hand. It facilitates mathematical verification. It makes possible extremely elaborate cross-checking. It can be designed for automatic generation of program code.

HOS with its axiomatic verification would be far too difficult to be a good manual technique. It *needs* automation. Its graphics tool is not merely cosmetic but is performing elaborate computation while the designer sits at the screen. The information engineering tools that are now emerging anchor the designer to centrally maintained data models and encyclopedias, and in some cases perform program generation. Most information engineering techniques do not work well in practice without automation. They require power-tool graphics.

Once the concept is accepted that systems analysts will employ graphics workstations, new techniques can be devised that are highly powerful but which would not work well if done by hand. The challenge in computer methodologies today is to employ such tools to advance the state of the art.

Figure 23.3 lists the same diagramming techniques as Fig. 23.1 and indicates properties other than properties concerned with their drawing capabilities. Let us briefly summarize our views on these techniques:

1. Decomposition Diagrams (Fig. 24.1)

- A simple means of diagramming the structure of organizations and complex processes.
- End users easily understand and draw decomposition tree structures.

2. Data Flow Diagrams (Fig. 24.4)

- An essential and valuable tool for understanding the flows of documents and data among processes.

OTHER PROPERTIES OF DIAGRAMMING TECHHNIQUES	DECOMPOSITION DIAGRAMS	DEPENDENCY DIAGRAMS	DATA FLOW DIAGRAMS	ENTITY-RELATIONSHIP DIAGRAMS	DATA STRUCTURE DIAGRAMS	DATA NAVIGATION DIAGRAMS	HIPO CHARTS	STRUCTURE CHARTS	WARNIER-ORR CHARTS	MICHAEL JACKSON CHARTS	FLOWCHARTS	STRUCTURED ENGLISH	NASSI-SHNEIDERMAN CHARTS	ACTION DIAGRAMS	DATA-BASE ACTION DIAGRAMS	DECISION TREES AND TABLES	STATE-TRANSITION DIAGRAMS	HOS CHARTS
Easy to read	YES	YES	YES	YES	YES	YES	X	YES	YES		o		YES	YES	YES	YES		
Quick to draw and change	YES	YES	YES	YES	YES	YES	X		YES					YES	YES	YES		
User-friendly (easy to teach to end users)	YES	YES	YES	YES	YES	YES	X	Fair	YES				Fair	YES	YES	YES		
Good for stepwise refinement							X	YES	YES					YES	YES			YES
One tool can show high-level structures and detailed logic														YES	YES			YES
Ultimate decomposition (can be decomposed to executable code)														YES	YES			YES
Good for computerized screen editing	YES	YES	YES	YES	YES	YES		YES	YES	YES				YES	YES	YES	YES	YES
Can be printed out on normal-width paper (without excessive division into pieces)				YES	YES	YES			√		YES			YES	YES	YES	YES	YES
Automatically convertible to program skeleton		YES				YES		YES	YES	YES			YES	YES	YES		YES	
Automatically convertible to executable program code														YES (4GL)	YES (4GL)	YES (Code Module)		YES
Problem-related rather than program-related terminology	YES	YES	YES	YES	YES											YES	YES	
Computerized accuracy checking														(can check uses of date)		YES		YES
Rigorous (with computerized checking)																		YES
Good for large, complex specifications which (with computerized assistance) need to be as error free as possible		YES			YES	YES								YES	YES	YES		YES
Oriented to interactive data-base usage and data modeling				YES	YES	YES									YES	YES		

X = High level only   o = Small charts only   √ = Up to a point

Figure 23.3 Other properties of diagramming techniques.

- End users can be quickly taught to read, check, and sometimes to draw data flow diagrams.

- Automated tools for drawing and manipulating data flow diagrams exist.

- While data flow diagrams are an excellent overview tool, they are not good for drawing program architectures. They have been overused and misused in this more detailed area.

- Use of multiple-layered data flow diagrams for creating complex specifications often results in bad specifications with inconsistencies, omissions, and ambiguities. Data flow diagrams must be linked to other tools.

- Data flow diagrams should be tightly linked to data models; otherwise, they can give false representations of data. Often this is not done.

- In some cases data flow diagrams need improving to show synchronization among separate events.

- Data flow diagrams need improving to show data layering as described in Chapter 7.

- Tools other than data flow diagrams are essential for diagramming the structures of processes and these should be designed for maximum verification. They could be linked to data flow diagrams.

3. Dependency Diagrams (Fig. 24.2)

- A replacement for data flow diagrams with similar ability to represent the flow of data among processes, but designed to be automatically convertible to action diagrams.

- Uses notation to represent optionality, conditions, cardinality, mutual inclusivity, and mutual exclusivity (which data flow diagrams do not conventionally use).

- Needs linking to soundly designed data models.

- A valuable component of a CASA/CAP graphics tool kit with links to other tools.

4. Entity-Relationship Diagrams (Fig. 24.5)

- An essential tool.

- Provide a logical overview of the data needed for running an operation (project, department, division, or entire enterprise).

- An essential part of strategic data planning.

- End users can quickly be taught to read, check, and sometimes to draw entity-relationship diagrams.

- An essential component of a CASA/CAP workbench which uses the sequence

5. Data Structure Diagrams (Fig. 24.12)

- An essential tool.

- An expansion of an entity-relationship diagram into detail showing data items.

- A means of representing correctly normalized data.

- All systems analysis should employ data analysis and data modeling at an early stage. Correctly normalized data structures should be designed prior to any procedure-level development or programming.

- A means of representing user views of data which are synthesized by computer into an overall, correctly normalized data model.

- Should identify formulas or algorithms for derived data items.

- End users can be taught to read and check data structure charts and sometimes to draw bubble charts representing end-user views of data.

- Bachman notation cannot represent functional dependencies in data and should not be used. The notation in Chapters 19 and 20 is preferable.

- Automated tools should be used for data modeling, normalization, and presenting analysts or programmers with subsets of data models.

- An essential component of a CASA/CAP workbench.

6. Data Navigation Diagrams (Fig. 24.6)

- A simple and essential tool for designing data-base navigation. Also useful with file systems.

- An adjunct to data modeling.

- An essential piece of documentation for data-base applications which ought to be an installation standard.

- Automatically convertible to action diagrams.

- Commonly hand-drawn on data submodel diagrams, but much better if drawn with software which performs automatic conversion to action diagrams.

- An essential component of a CASA/CAP workbench which uses the sequence

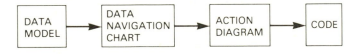

7. HIPO Charts (Fig. 10.2)

- We believe that HIPO diagrams should not be used because other methods are better.

- Data flow diagrams or dependency diagrams give a much more compact and easy-to-read view of the flows of data.

- A high-level HIPO diagram (i.e., visual table of contents) does not give complete control structure information.

  (a) It does not describe the control constructs and control variables governing module invocation.
  (b) It does not describe the input and output data for each procedural component.
  (c) It does not include a link to a data model or data dictionary.

- A detail-level HIPO diagram (i.e., overview and detail HIPO diagrams) is limited to defining procedural components.

  (a) It does not have symbols for representing the basic control constructs of sequence, selection, and repetition.
  (b) It does not include a link to a data model or data dictionary.

8. Structure Charts (Fig. 9.2)

- A commonly used technique for showing program hierarchy used in conjunction with data flow diagrams and pseudocode.

- The structure chart does not give complete control structure information.

  (a) It does not describe the control constructs and the control variables governing the invocation of procedural components. Some structure charts do include some of this information, but it is optional and not necessarily complete.
  (b) It does not generally show sequence, conditions, case structures, and loop control, and hence has to be supplemented with structured English or pseudocode.
  (c) It does not describe the input and output data for each procedural component.
  (d) It does not include a link to a data model, data dictionary, or charts used to define the program data structures.

- Structure charts can become messy when data and control variables are written on them, as in Fig. 9.2.

- Structure charts prevent rather than assist in automated verification.

- An early and rather inadequate way of diagramming program structures which ought to be replaced with better techniques.

9. Warnier–Orr Charts (Fig. 11.5)

- A user-friendly type of diagram for showing functional decomposition and hierarchical data structures.

- More compact than structure charts, HIPO charts, or Michael Jackson charts, but spread out horizontally more than action diagrams.

- Easy to read, draw, and change.

- They show sequence, selection, and repetition, which structure charts do not.

- They do not show the conditions or variables that control selection, case structures, or loop structures. (This information is relegated to footnotes.)

- They do not show input and output data for procedural components.

- They do not facilitate automated checking. They are species I functional decomposition.

- They provide no *direct* link to a data model or to a data dictionary. Also there is no link between Warnier–Orr diagrams representing procedural and data components for the same program.

- They are not data-base oriented. They can represent only hierarchical data structures.

- At a detail level they degenerate into a form of pseudocode which can be more difficult to read and draw than action diagrams. (Note the BEGIN-END blocks in Fig. 11.5.)

- While liking Warnier–Orr diagrams, we would not use them because action diagrams (including data-base action diagrams) have many advantages over them.

10. Michael Jackson Charts (Fig. 12.5)

- Michael Jackson methodology, described in Chapter 12, is claimed to give fewer errors in program structure than Warnier–Orr or Yourdon–Constantine methodologies. In installations we examined this appeared to be true, but we could find no firm statistics.

- Michael Jackson methodology designs the data first and derives the program structure from them. We find this a helpful and desirable approach.

- The methodology promotes consistent program designs.

- Perhaps the most difficult of the methodologies to learn and use correctly.

- Not user friendly.

- Oriented to file, not data-base operations.

- Represents only hierarchical data structures.

- Tends to break down or become very difficult to apply to complex programs. Some installations have abandoned Michael Jackson methodology for this reason.

- Shows sequence, selection, and repetition, which structure charts do not.

- Does not show the conditions or variables that control selection, case structures, or loop structures.

- To show detail design, they resort to a form of pseudocode (structured text) which is difficult to understand.

- IBM [1] commented that their development groups found it *"unsuitable in a highly volatile data-base/data-communication environment where data bases are very large and complex and user changes continually affect the structure and contents of the data bases. This type of decomposition forces programmers to discard a great amount of code whenever the data structures change, which is very expensive.*

- We prefer to use either action diagrams with fourth-generation languages (which is much easier) or HOS methodology (which is much more rigorous).

11. Flowcharts (Fig. 13.2)

- Not a structured technique.
- Tends to lead to unstructured code which is difficult to maintain.
- Should be avoided in favor of structured diagramming techniques.

12. Structured English and Pseudocode (Fig. 14.2)

- A picture is worth a thousand words, and this is not a pictorial technique.
- Sometimes too lengthy. Often longer than fourth-generation-language code.
- Some forms of pseudocode are highly mnemonic and difficult to read.
- Can be helpful to programmers.
- We would replace it with action diagrams, which at a detail level are somewhat similar but are pictorial.

13. Nassi–Shneiderman Charts (Fig. 15.2)

- Show detailed logic only, not program architecture or functional decomposition.
- Easy to read. Graphically appealing.
- Easy to teach to end users.
- Generally better than flowcharts or pseudocode, which they replace.
- Too time consuming to draw and change.
- Not linked to a data dictionary or data model.
- Can show neither high-level program structure nor low-level degeneration into code.
- We prefer action diagrams (and data-base action diagrams), which overcome the deficiencies noted above.

14. Action Diagrams (Fig. 16.6)

- A simple, elegant technique designed to overcome many of the deficiencies in earlier techniques.
- Quick to draw. Easy to computerize.
- Easy to read.
- Designed to be easy to teach to end users.
- A single technique that extends all the way from the highest-level functional decomposition to the lowest-level logic and coding. Can be decomposed into executable code. All other methods described so far require two or three different forms of representation to accomplish this.
- Can have the wording of a chosen fourth-generation language.

- Enforces correct control structures.
- Designed to spread out vertically, not horizontally, so that an inexpensive desktop printer can print the charts.
- Gives species II functional decomposition and so facilitates cross-checking of inputs and outputs.
- Action diagrams can show nonprocedural data-base operation (compound data-base actions).
- Quick and attractive for computerized editing.
- Program design can be taken to its final state in a language-independent fashion and then, largely automatically, converted to the language in question.
- Decomposition diagrams, dependency diagrams, data navigation charts, and decision trees can be automatically converted to action diagrams.
- A highly useful component of a CASA/CAP workbench.

15. Decision Trees and Tables (Fig. 24.13)

- A valuable technique for representing complex sets of conditions or rules and the actions that result from them.
- Software exists to convert decision tables into program modules.
- Every analyst and programmer should be familiar with them. They are an important tool in a DP professional's tool kit.
- Users can be taught to check decision tables for complex sets of rules or conditions.

16. State-Transition Diagrams (Fig. 24.3)

- A specialized technique valuable for certain types of complex logic where multiple transitions among states can occur.
- Systems analysts should understand the technique so that they can apply it if appropriate.

17. HOS Charts (Fig. 8.6)

- Species III functional decomposition.
- More rigorous than any other structured technique.
- Mathematically based so that designs which are provably correct are created.
- Automatically convertible to correct program code.
- Not user friendly.
- A tool for a professional analyst.
- Requires a commitment to learn a technique substantially different from traditional techniques.

- Extends all the way from the highest-level functional decomposition to automatic generation of program code.

- Eliminates most debugging.

- Extremely valuable for creating highly complex specifications where any other method would give errors, inconsistencies, omissions, and ambiguities.

- Eventually, the entire world of structured analysis and programming must move to tools that have this level of rigor and automation.

WHAT TECHNIQUES WOULD WE USE?   Now for the $64,000 question. Which of these various techniques would we select for running a DP operation?

The left-hand side of Fig. 23.4 is fairly clear. We need entity-relationship diagrams and data structure diagrams. The right-hand side offers more choice.

Either HOS methodology or action diagrams give a single technique which is applicable from top to bottom. HOS is the most rigorous of any structured technique. It gives designs which are guaranteed to be internally consistent and correct, and it automatically generates correct code from them. The designer builds his system with graphics at a computer terminal, and at this terminal the system is easy to modify and maintain.

If the system in question is highly complex, like military systems or the control of complex operations, it is asking for trouble to use methods which do not have the automated verification that HOS has. Experience has shown over and over again what an unholy mess results from the use of traditional structured methods and even worse from nonstructured methods.

The problem with HOS is that it is different from conventional methods. The analysts need training and will take some time to become skilled with it.

In a less complex commercial DP environment, we would employ the most appropriate fourth-generation languages, report generators, and so on, and their associated data management systems. Action diagrams are designed for this environment and work well with traditional languages. They are easy to teach to end users and encourage end users to progress from understanding high-level diagrams of the organization chart and functional decomposition down to diagrams of program detail drawn with the same technique.

Decision tables are a valuable tool for handling complex conditions and actions such as those in Fig. 17.3. Analysts should understand this tool and use it when appropriate, preferably with automated code generation.

Figure 23.5 shows our preferences. We would want all analysts in a DP organization to be thoroughly trained to use:

- Decomposition diagrams.
- Dependency diagrams (including data flow diagrams)

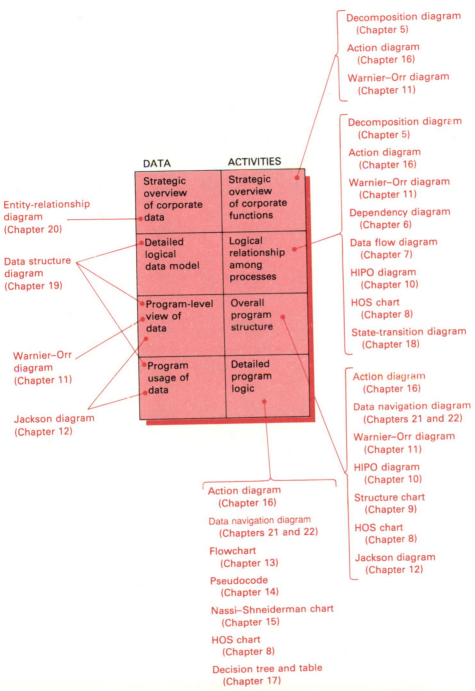

Decomposition diagram
(Chapter 5)

Action diagram
(Chapter 16)

Warnier–Orr diagram
(Chapter 11)

Decomposition diagram
(Chapter 5)

Action diagram
(Chapter 16)

Warnier–Orr diagram
(Chapter 11)

Dependency diagram
(Chapter 6)

Data flow diagram
(Chapter 7)

HIPO diagram
(Chapter 10)

HOS chart
(Chapter 8)

State-transition diagram
(Chapter 18)

Entity-relationship
diagram
(Chapter 20)

Data structure
diagram
(Chapter 19)

DATA   ACTIVITIES

Strategic overview of corporate data	Strategic overview of corporate functions
Detailed logical data model	Logical relationship among processes
Program-level view of data	Overall program structure
Program usage of data	Detailed program logic

Warnier–Orr
diagram
(Chapter 11)

Jackson diagram
(Chapter 12)

Action diagram
(Chapter 16)

Data navigation diagram
(Chapters 21 and 22)

Warnier–Orr diagram
(Chapter 11)

HIPO diagram
(Chapter 10)

Structure chart
(Chapter 9)

HOS chart
(Chapter 8)

Jackson diagram
(Chapter 12)

Action diagram
(Chapter 16)

Data navigation diagram
(Chapters 21 and 22)

Flowchart
(Chapter 13)

Pseudocode
(Chapter 14)

Nassi–Shneiderman chart
(Chapter 15)

HOS chart
(Chapter 8)

Decision tree and table
(Chapter 17)

Figure 23.4   Areas in which different diagramming techniques are applicable.

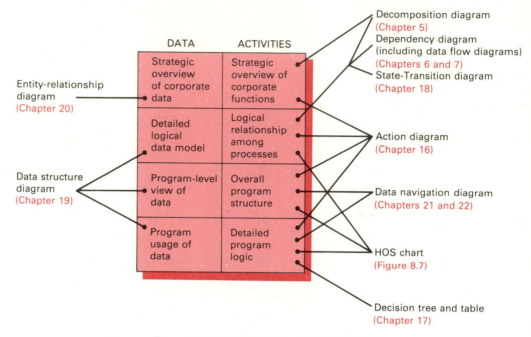

Figure 23.5   Choice of diagramming techniques.

- Action diagrams
- Entity-relationship diagrams
- Data structure diagrams
- Data navigation diagrams
- Decision trees and tables

Some analysts should also be taught to use state-transition diagrams. Where HOS is selected as a methodology because of its ability to create error-free specifications, the analysts should be taught this diagramming technique.

CHALLENGE

The challenge of the future is to adapt these drawing techniques and the software that goes with them into a systems engineering tool kit which is easy and fast to use, and provides the most rigorous checks possible on the emerging design. This tool kit should generate executable code from the detailed diagrams. The diagrams should be easy to edit and change because complex designs change and evolve constantly during their creation. It should be possible to change the diagrams while enforcing rigor, so that maintenance can be performed quickly.

It is possible to have an analyst's workbench that generates bug-free programs and specifications which are internally consistent and correct. The challenge today is to adapt this rigor to techniques that are user friendly and as powerful as possible.

Computerized tools for the analyst are essential. The tools should automate the process of data modeling and third-normal-form design [2], automatic extraction of views of data from the models, the building of data navigation diagrams on the data models, and automatic conversion of these to action diagrams. The tool should support the creation and modification of decomposition diagrams and dependency diagrams showing data flow and the automatic conversion of these to action diagrams. It should facilitate the editing of action diagrams and the conversion of them to executable code. This process may support various fourth-generation languages as well as third-generation languages.

Computerized tools should provide the maximum possible cross-checking and verification of the design, and verification that data are designed and used correctly. Species II (or III) for decomposition is needed with computerized verification that inputs and outputs are consistent.

Figure 23.6 shows a basic short list of the diagramming techniques and how they interlink. Automatic or computer-aided conversion is shown by the heavy

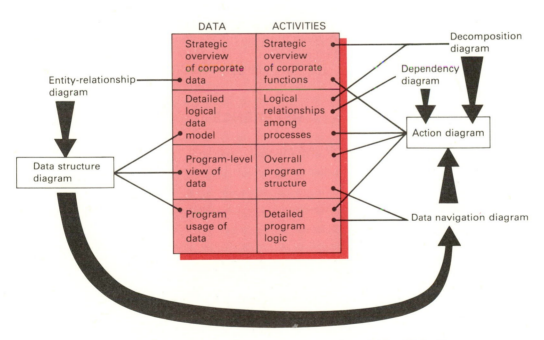

Figure 23.6   Basic short list of the diagramming techniques in Fig. 23.5 with automatic or computer-aided conversion indicated by the heavy arrows. Executable code should be generated from the action diagram and data structure diagram, as shown in Fig. 23.7.

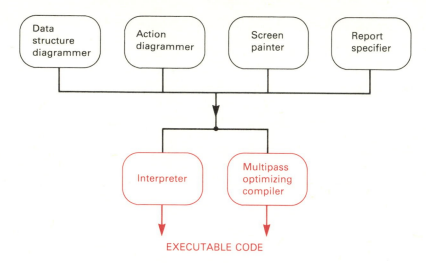

Figure 23.7   The data structure diagrams and action diagrams of Fig. 23.7 should be used to generate executable code.

arrows. From the data structure diagram and action diagram, executable code should be generated. Code generators require specifications of screens and reports that are used. Certain fourth-generation languages contain excellent screen pointers and report specifiers. Figure 23.7 indicates how these, combined with diagramming techniques, can create a graphically oriented application generator.

REFERENCE

1. G. R. Rogers, A Simple Architecture for Consistent Application Program Design, *IBM Systems Journal,* IBM, Armonk, NY, Vol. 22, No. 3, 1983.

24 A RECOMMENDED SET OF DIAGRAMMING STANDARDS

> *Note:* This chapter is a summary which repeats information from previous chapters.

INTRODUCTION
It is desirable to provide systems analysts with a set of diagramming techniques with which they can conceptualize, analyze, and design. These techniques should act as an aid to clear thinking. It is desirable that the diagramming conventions be consistent across the set of graphic tools. In the future these graphic tools will be employed at the screen of a computer. Computer-aided design will greatly speed up and improve the quality of the analyst's and designer's work.

This chapter presents an integrated set of diagramming conventions. They tie together and unify many of the best ideas of the earlier chapters. We believe that these should be taught to every systems analyst and designer as part of his training. Creating diagramming standards is desirable in any corporation because it greatly aids communication among DP professionals and between DP and end users.

BASIC GRAPHIC TOOLS
The analyst should understand and be able to use the following set of diagramming techniques:

- Decomposition diagrams
- Dependency diagrams

- Data flow diagrams
- Data structure diagrams
- Entity-relationship diagrams
- Data navigation diagrams
- Action diagrams
- Decision trees and tables
- State-transition diagrams and tables

Figure 24.1 to 24.13 at the end of the chapter show examples of these diagrams with a consistent notation printed from a computer screen.

The list above provides a basis for the analyst's education and a basis for complete methodologies for analysis and design. The rest of this chapter describes a recommended set of diagramming standards. The same conventions are used on all of the types of diagrams listed above.

SQUARE AND ROUND CORNERS   It is important to distinguish carefully between data and processes on diagrams. To do so, data-entity types or record types are drawn as square-cornerd boxes; functions, processes, procedures, or activities in general are drawn as round-cornered boxes. Thus:

DATA

ACTIVITIES

Circles, ellipses, and other shapes are used for anything that is not a data-entity type or record type or a function or process or procedure. For example, ellipses are used to represent data-item types in a bubble chart for data analysis; circles are used to show states in a state-transition diagram.

LEVELS OF DETAIL   Design, as indicated earlier, occurs at differing levels of detail. The highest is the strategic or overview level; the next is the level of logical structures; the lowest is the level of physical or implementation structures. A computer can use different degrees of shading to represent these levels, thus:

	DATA	ACTIVITIES
Overview level	ROUGH ENTITY TYPE	FUNCTION
Logical analysis or design	NORMALIZED ENTITY TYPE	PROCESS
Implementation level	PHYSICAL RECORD TYPE	PROCEDURE

CARDINALITY

An instance of block A may be associated with N instances of block B. A cardinality symbol is placed next to block B on the line from A to B. It gives information about the possible values of N.

A 1-with-many association is indicated by a crow's foot.

A 1-with-1 association is indicated with a bar (like a small "1").

If there may be zero instances of block B associated with block A, a zero is shown next to the "1" bar or crow's foot.

A minimum and maximum value may be associated with the crow's foot.

These conventions apply to either data boxes or activity boxes:

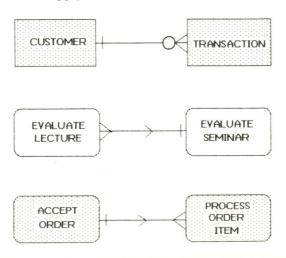

Analysts often draw a line with no cardinality mark on it to mean a one-to-one association. On entity-relationship diagrams it is recommended that the one-to-one indicators always be drawn. A relationship without a mark often means that the cardinality is not known or has not been thought about. To ensure completeness of the entity-relationship diagram a computerized tool should insist that cardinality be stated at the end of each relationship line. If the analyst does not yet know the cardinality, he may put a question mark on the end of the line.

FLOW AND TIME SEQUENCE   Lines between round-cornered (activity) boxes need an arrow to show flow or sequence. Dependency diagrams and flow diagrams can thus have lines with both cardinality indicators and arrows.

HORIZONTAL AND VERTICAL LINES   Lines between boxes should be drawn horizontally and vertically. This assists with the labeling conventions, avoids distorted crow's-feet, and enables the computer to employ a palette of symbols which do not need to be rotated.

Diagrams containing hierarchies should be drawn with root boxes at the top and left, where possible. One-to-many lines should go downward and to the right where possible. Hierarchical patterns in data become familiar to the analyst as having the following type of shape:

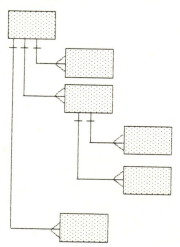

Computerized tools used to synthesize complex data models can draw them in this way automatically.

When drawing a data navigation chart on top of an entity-relationship diagram, the analyst may rearrange the diagram for maximum clarity.

On diagrams with activity boxes, the arrows representing flows or time-sequences should go downward or left to right where possible.

LABELING OF LINES   The lines connecting boxes may be labeled. A line on an entity-relationship diagram may be labeled with the name of the relationship.

The label *above* a horizontal line is the name when the diagram is read from left to right.

The label *below* a horizontal line is the name when the diagram is read from right to left.

If the line is vertical the label on the left is read going *down* the line. The label on the right is read going *up* the line.

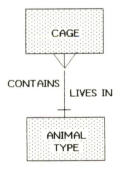

OPTIONALITY   A dot in a broken line means optionality. Whatever follows the dot may or may not exist.

A single sentence may be used to describe the lines on the diagrams. The dot at the start of a line is read as "sometimes". The dot at the end of the line is ignored. (It is read when traversing the line in the opposite direction.)

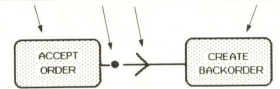

ACCEPT ORDER SOMETIMES IS FOLLOWED BY CREATE BACKORDER

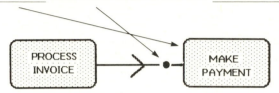

MAKE PAYMENT SOMETIMES OCCURS WITHOUT PROCESS INVOICE

A dot, meaning "sometimes," is equivalent to a zero in the cardinality symbol. In entity-relationship diagrams the zero is normally used as part of the symbol sng maximum and minimum cardinality values. The dot is then not needed.

The dot, like crow's-feet and other symbols, can be used on decomposition diagrams, dependency diagrams, or data flow diagrams. A condition written against a dot can be transferred to the equivalent action diagram:

DATA NAVIGATION DIAGRAM   EQUIVALENT ACTION DIAGRAM

IF ORDER-QUANTITY > 50 ●   IF ORDER-QUANTITY > 50

READ INVENTORY   READ INVENTORY

MUTUAL EXCLUSIVITY   It is often necessary to indicate that one and only one of several choices will be used. In other words, the choices will be mutually exclusive. This is indicated by a large dot with a branch on a line, thus:

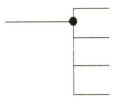

This can be used on decomposition diagrams, entity-relationship diagrams, data navigation diagrams, decision trees, data flow diagrams, dependency diagrams, or state-transition diagrams.

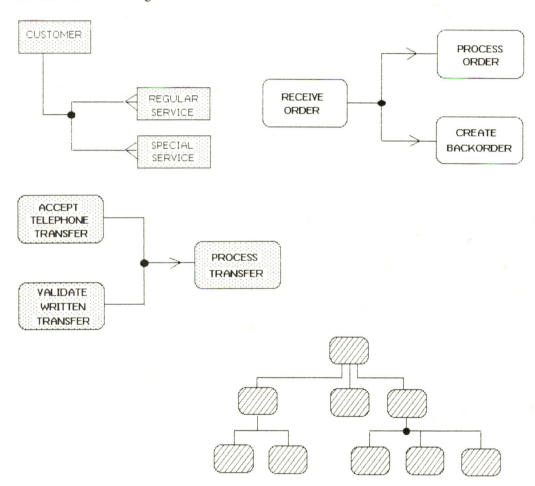

A tree structure like that above is often drawn with branching lines. The dot on the branch shows mutual exclusivity. The analyst may think of the dot as being an ''o'' for ''or.''

Mutual exclusivity in action diagrams (or programs) is handled with a case structure. This is drawn on action diagrams with a similar shape to the branching line:

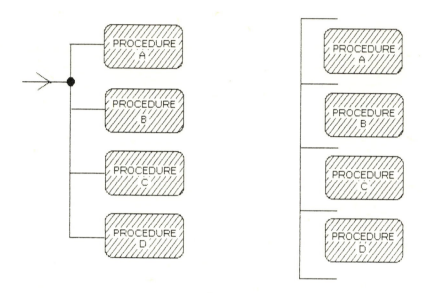

In the choice of two lines A or B, the following possibilities exist:

Dependency, flow, relationship or navigation diagram:	Action diagram equivents:
A and B: ![A B]	A B or B A
A or B: ![A B]	A B
A and B or neither: ![A B CONDITION]	CONDITION A B or CONDITION B A

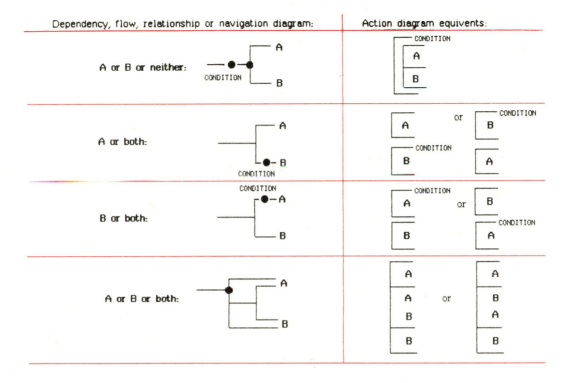

CONDITIONS

The dot on a diagram, representing optionality, is associated with a condition. A number or name on the diagram may refer to a separate list of conditions. Where possible, however, it is clearer for the reader if the condition is written on the diagram:

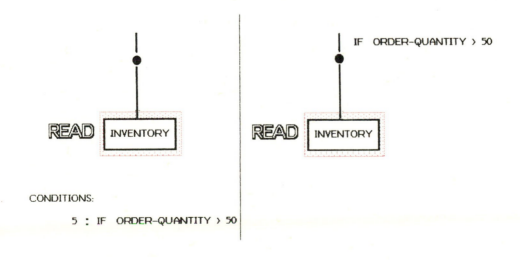

A mutually exclusive indicator may also have conditions associated with it:

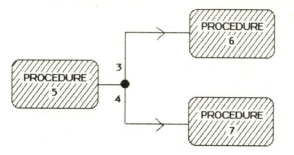

CONDITIONS:

3 : IF OPERATOR ENTERS " YES"

4 : IF OPERATOR ENTERS " NO"

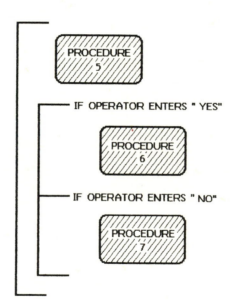

DATA STRUCTURE DIAGRAMS

Data analysis is concerned with functional dependencies among data items. A functional dependency is a one-to-one association; for example, for one instance of data item A there is one and only one instance of data item B. It is shown with a "one-to-one" bar across the link:

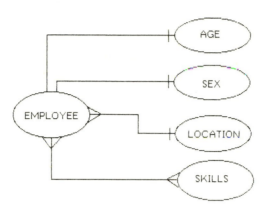

The crow's-foot has its usual meaning of one-to-many association.

In the diagram above, there is no mark on the lines *from* AGE to EMPLOYEE, or *from* SEX to EMPLOYEE, because these are not of interest in data analysis. The normalization algorithm is interested in all one-to-one associations. It treats anything not marked as though it is *not* a one-to-one association.

Data analysis arrows can also be drawn on box-shaped record layouts, thus:

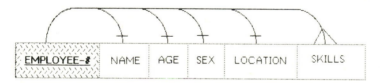

DERIVED DATA ITEMS

A data analysis chart may show a derived data item, with the formula or algorithm for deriving it. To do this, a dashed line, arrows and crow's-feet are used:

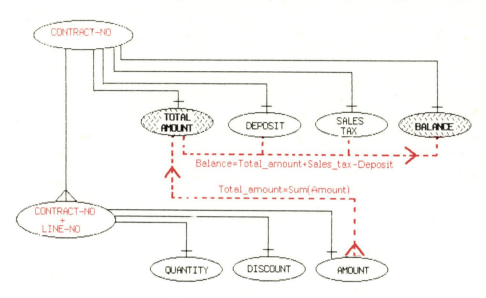

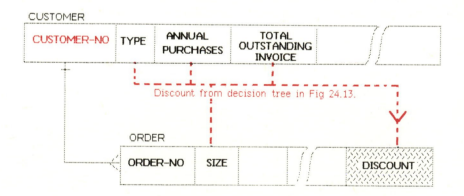

ENTITY SUBTYPES

Entity subtypes, when used, are shown as subdivisions of entity types:

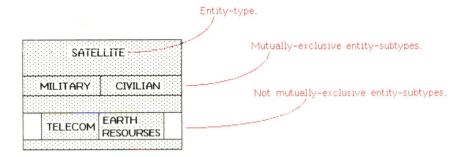

Associations to other entity types may be drawn from the subtypes:

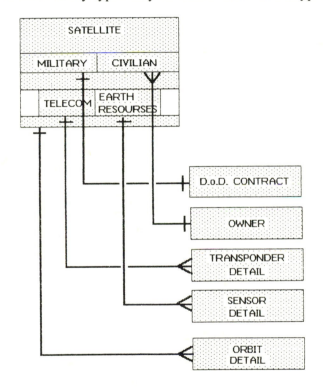

SEQUENCE   High-level decomposition diagrams are usually un-
concerned with sequence. They use a tree structure to
show how a function is composed of lower-level functions. Lower-level decomposition diagrams may need to show sequence. They may show how a process

is composed of subprocesses which are executed *in a given sequence*. To show this, an arrow is used pointing in the direction of the sequence. This direction should be drawn top to bottom on a horizontal tree or left to right on a vertical tree.

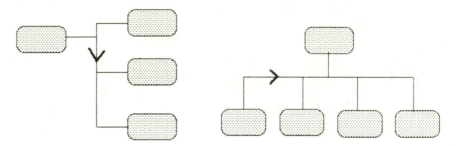

A thin line is to show the flow of computer data; a thick line is used to show the flow of physical documents or materials:

DATA NAVIGATION DIAGRAMS

A chart showing navigation through data may be drawn on top of the appropriate data model.

Simple accesses are drawn as a single rectangle:

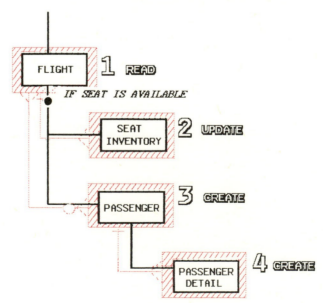

Compound accesses are shown with a double box:

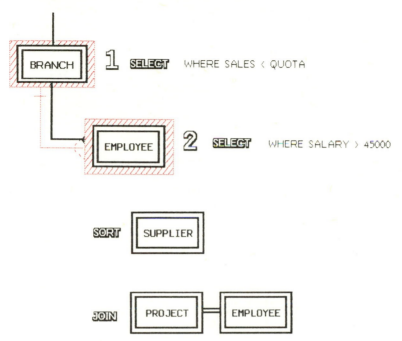

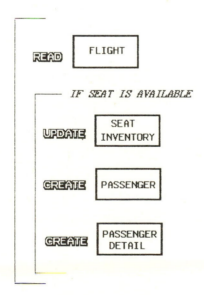

The access symbols on the navigation chart can be transposed directly and automatically to an action diagram:

EVENTS

A large arrow is used on a diagram to show that an event occurs:

This may be used on a dependency diagram, data flow diagram, or state-transition diagram.

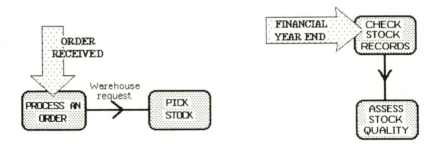

CONNECTORS

A triangle is used as a connector, to connect lines to a distant part of a diagram:

The triangle may be used to connect to other pages. This is often unnecessary with computerized graphics because the user scrolls across large, complete diagrams.

NESTING

To make them easier to understand, large diagrams should be nested. A block can be enlarged into a diagram which shows the contents of the block. This diagram should be surrounded by a frame of the same shape as its parent block.

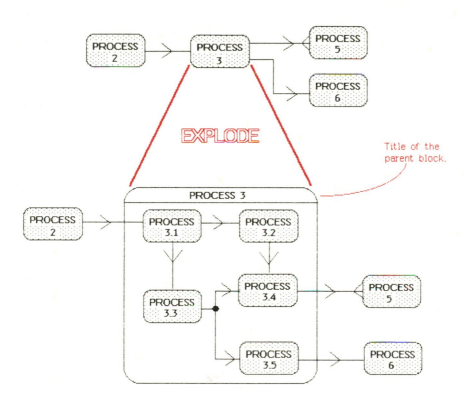

Using computer graphics, the analyst may point to any block and ask to see it in more detail. Conversely, he may put a box around any group of blocks and shrink them to one block. The process of enlarging as in the diagram above is called *exploding*. The converse process is called *imploding*. The menus with which the analyst manipulates graphics may contain the words EXPLODE and IMPLODE.

OTHER FORMS OF EXPANDING AND SHRINKING

The terms EXPLODE and IMPLODE refer to nesting in which the type of diagram remains the same. A block may also be expanded in other ways to show *different kinds* of detail.

Box 16.1 shows commands and subcommands for exploring large diagrams.

Specific types of detail may be referred to with specific names such as ATTRIBUTES, SUBTYPES, NAVIGATION DIAGRAM, and so on.

BOX 16.1   Commands and subcommands for computer manipulation of large diagrams

PAGE: LEFT, RIGHT, UP, DOWN, JUMP
  Reveals a different page of a large diagram.
SCROLL: LEFT, RIGHT, UP, DOWN
  Moves in variable increments across a diagram.
ZOOM: IN, OUT
  Expands or shrinks the diagram without changing it.
ZOOC: IN, OUT
  Zoom and change, or zoom and clarify: expands or shrinks the diagram like a zoom but adjusts it for readability or clarification.
NEST: EXPLODE, IMPLODE
  Changes to a more detailed or a less detailed diagram *of the same form*, for example revealing blocks within blocks.
DETAIL: SHOW, HIDE
  Changes to a more detailed or a less detailed diagram *of a different form*.
ANNOTATE: ADD, DELETE
  Adds or removes descriptive matter.

In the following illustration the arrow indicating a derived data item is expanded (detailed) to show a decision tree:

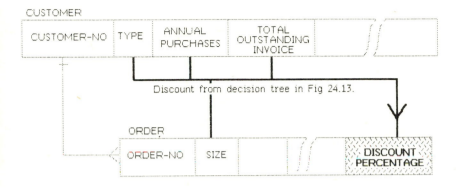

Discount from decision tree in Fig 24.13.

SHOW DETAIL

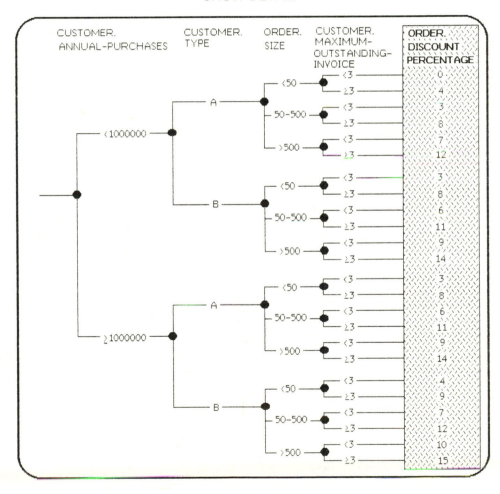

CROSSING LINES

When one line crosses another on a diagram without any logical linkage, one of the lines in question will be broken at the crossover:

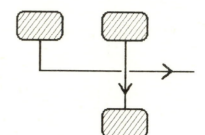

USE OF COLOR

When an analyst can employ color on a computer graphics facility, he is tempted to splash colors everywhere. Color can be a very powerful aid if it is used with *constraint*.

It should be used only to improve the understanding of the diagrams. It can do this in the following way:

- It can highlight items of special importance.
- It can highlight items to which the attention of the user should be directed.
- It can be used to separate comments or explanatory pointers from the substance of a diagram.
- It can distinguish between blocks or areas of different types.
- It can be used to simplify a complex diagram; the user looks at one color at a time.
- It can be used to overlay one type of drawing on another.
- It can have mnemonic value; different colors can be associated with different subjects.
- It can represent an extra dimension on charts with complex data.

These uses of color are very valuable, but color loses its value if it is used for purposes of decoration rather than logic. The instinct to be artistic must be suppressed and replaced with an urge to maximize clarity.

Where a monochrome display or printout is used, the effect of two colors may be simulated by making one image dotted or gray and the other image heavy black, as on the earlier data navigation charts, where the data structure is gray and the navigation path is drawn on top of it in black. Computer-generated shadings can also simulate colors, thus:

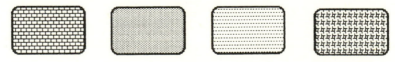

EXAMPLES

The remaining illustrations in this chapter show examples of the main types of diagrams that a systems analyst should use.

The PALETTE on each diagram shows the types of symbols used. Symbols may be taken from the PALETTE on the screen of a computer-aided design tool.

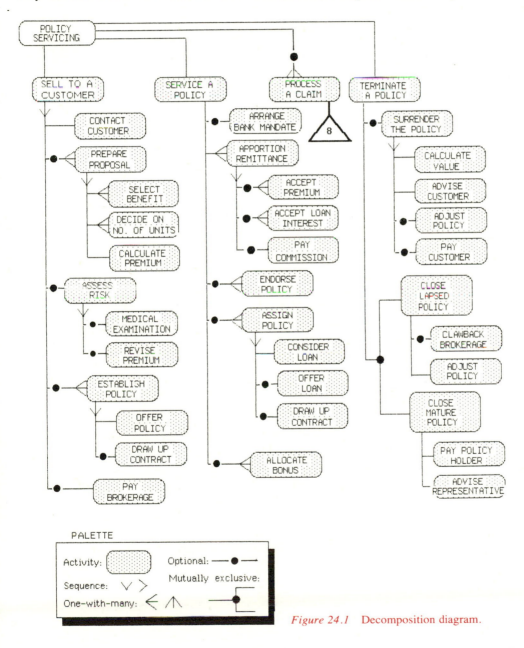

Figure 24.1 Decomposition diagram.

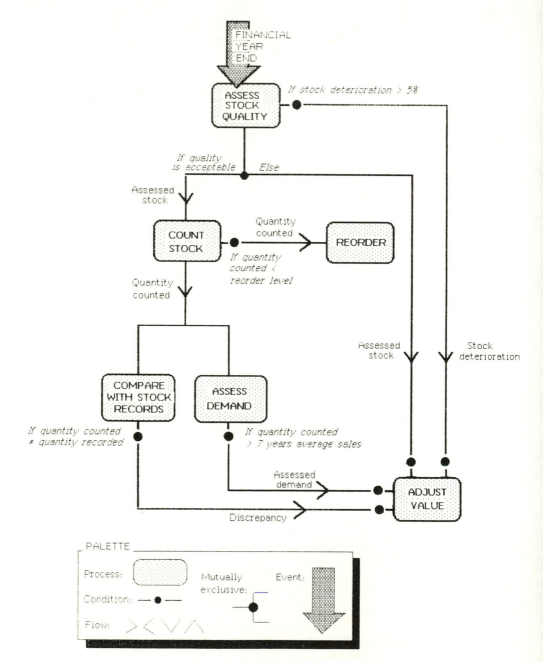

Figure 24.2   Dependency diagram.

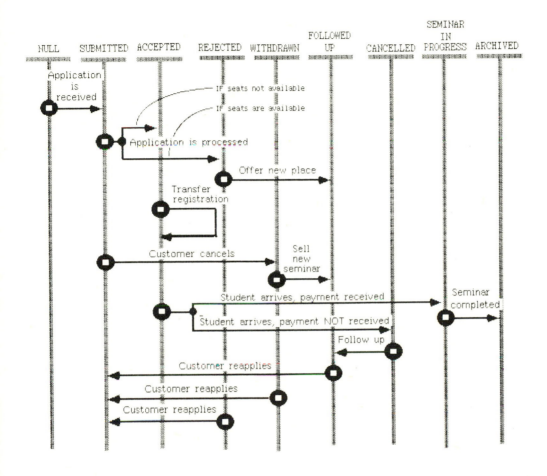

Figure 24.3 State-transition diagram.

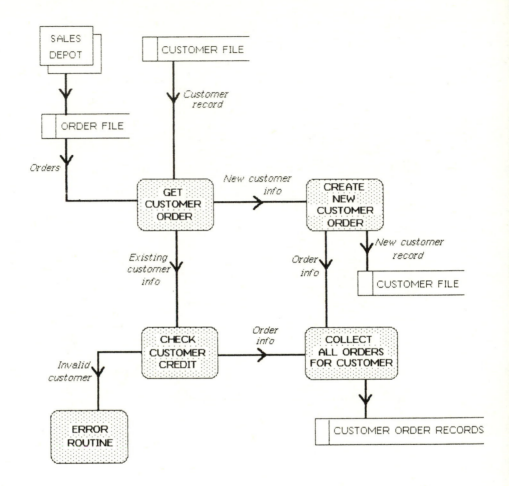

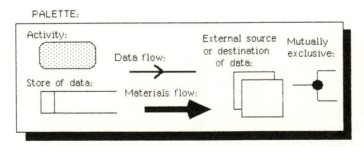

Figure 24.4 Data flow diagram.

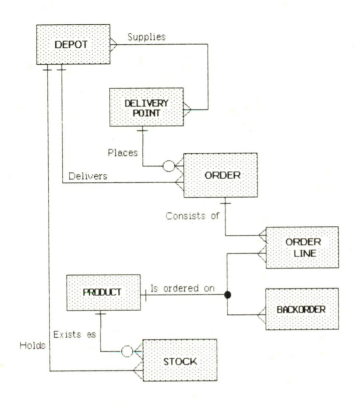

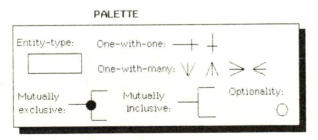

Figure 24.5 Entity-relationship diagram.

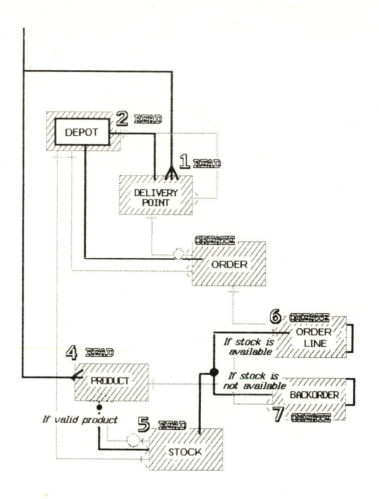

PALETTE

Simple operations:	Simple access:	Compound access:	Mutually exclusive:

CREATE

READ

UPDATE

DELETE

Cardinality: ⋁ ⋀

Condition: ●—

Figure 24.6 Data navigation diagram drawn on top of previous entity-relationship diagram.

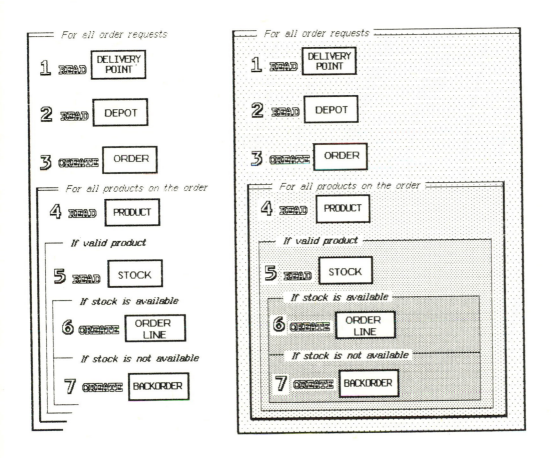

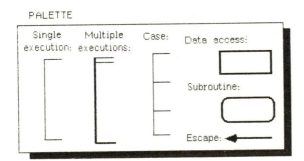

Figure 24.7 Action diagram, in alternate formats, derived from the data navigation diagram of Fig. 24.6.

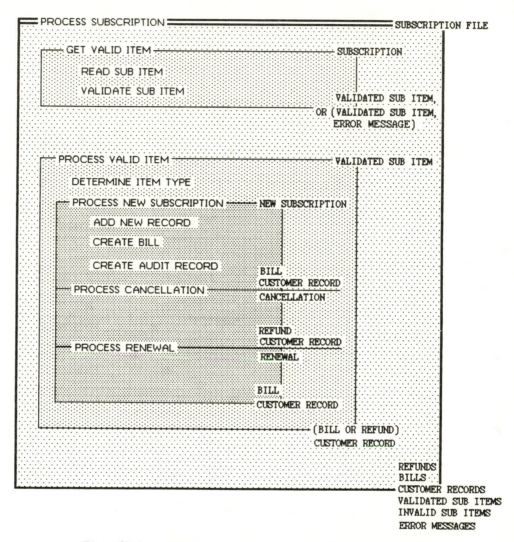

Figure 24.8 Action diagram showing subroutine inputs and outputs.

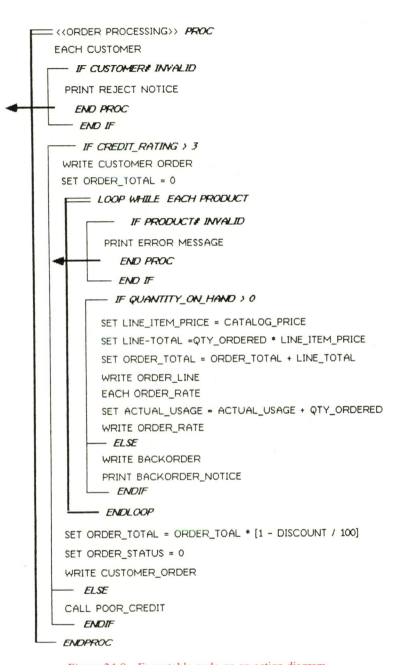

```
<<ORDER PROCESSING>> PROC
  EACH CUSTOMER
    IF CUSTOMER# INVALID
      PRINT REJECT NOTICE
      END PROC
    END IF

    IF CREDIT_RATING > 3
      WRITE CUSTOMER ORDER
      SET ORDER_TOTAL = 0
        LOOP WHILE EACH PRODUCT
            IF PRODUCT# INVALID
              PRINT ERROR MESSAGE
              END PROC
            END IF
            IF QUANTITY_ON_HAND > 0
              SET LINE_ITEM_PRICE = CATALOG_PRICE
              SET LINE-TOTAL =QTY_ORDERED * LINE_ITEM_PRICE
              SET ORDER_TOTAL = ORDER_TOTAL + LINE_TOTAL
              WRITE ORDER_LINE
              EACH ORDER_RATE
              SET ACTUAL_USAGE = ACTUAL_USAGE + QTY_ORDERED
              WRITE ORDER_RATE
            ELSE
              WRITE BACKORDER
              PRINT BACKORDER_NOTICE
            ENDIF
        ENDLOOP
      SET ORDER_TOTAL = ORDER_TOAL * [1 - DISCOUNT / 100]
      SET ORDER_STATUS = 0
      WRITE CUSTOMER_ORDER
    ELSE
      CALL POOR_CREDIT
    ENDIF
  ENDPROC
```

Figure 24.9 Executable code on an action diagram.

377

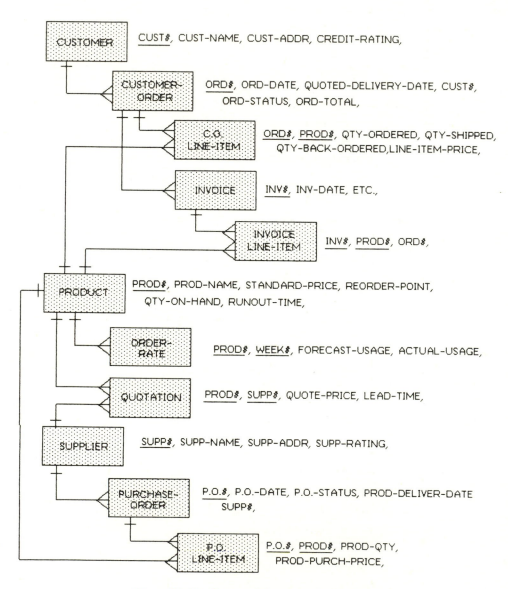

CUSTOMER — CUST#, CUST-NAME, CUST-ADDR, CREDIT-RATING,

CUSTOMER-ORDER — ORD#, ORD-DATE, QUOTED-DELIVERY-DATE, CUST#, ORD-STATUS, ORD-TOTAL,

C.O. LINE-ITEM — ORD#, PROD#, QTY-ORDERED, QTY-SHIPPED, QTY-BACK-ORDERED, LINE-ITEM-PRICE,

INVOICE — INV#, INV-DATE, ETC.,

INVOICE LINE-ITEM — INV#, PROD#, ORD#,

PRODUCT — PROD#, PROD-NAME, STANDARD-PRICE, REORDER-POINT, QTY-ON-HAND, RUNOUT-TIME,

ORDER-RATE — PROD#, WEEK#, FORECAST-USAGE, ACTUAL-USAGE,

QUOTATION — PROD#, SUPP#, QUOTE-PRICE, LEAD-TIME,

SUPPLIER — SUPP#, SUPP-NAME, SUPP-ADDR, SUPP-RATING,

PURCHASE-ORDER — P.O.#, P.O.-DATE, P.O.-STATUS, PROD-DELIVER-DATE SUPP#,

P.O. LINE-ITEM — P.O.#, PROD#, PROD-QTY, PROD-PURCH-PRICE,

Figure 24.10 Data model annotated with fields.

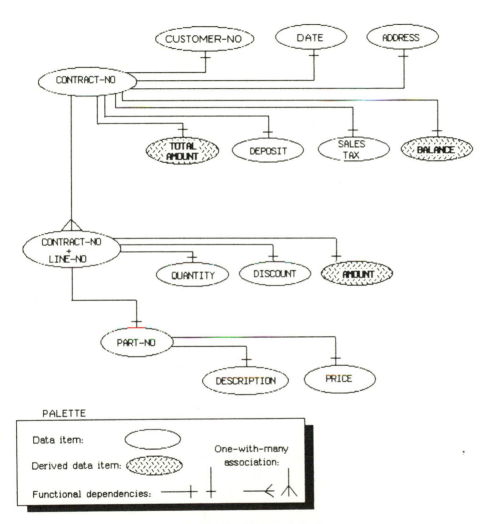

Figure 24.11 Data analysis diagram.

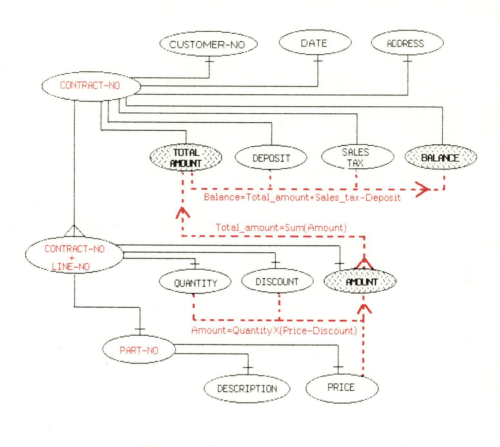

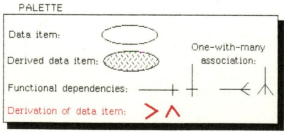

Figure 24.12   Data analysis diagram showing derivation.

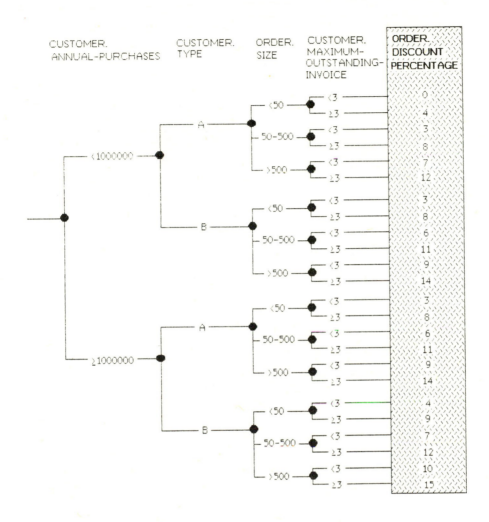

CUSTOMER.
ANNUAL-PURCHASES CUSTOMER. TYPE ORDER. SIZE CUSTOMER. MAXIMUM-OUTSTANDING-INVOICE ORDER. DISCOUNT PERCENTAGE

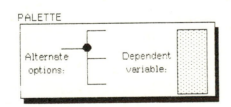

PALETTE

Alternate options: Dependent variable:

Figure 24.13 Decision tree showing the derivation of ORDER__DISCOUNT __PERCENTAGE.

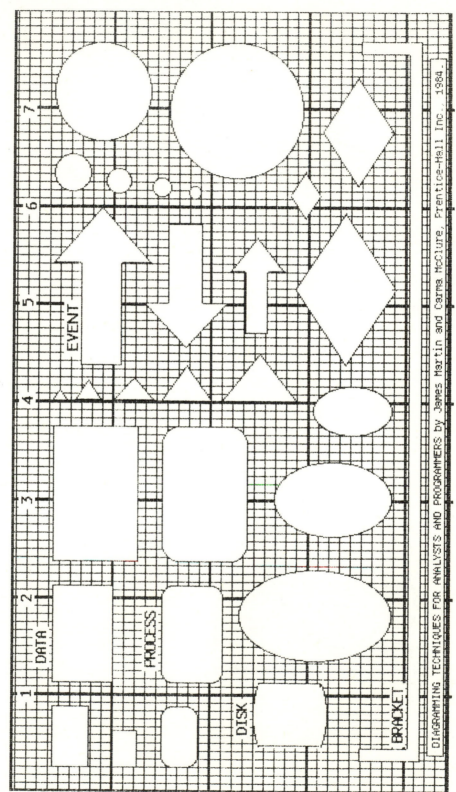

Figure 24.14 Template for creating the diagrams in this chapter.

INDEX

I

J

L

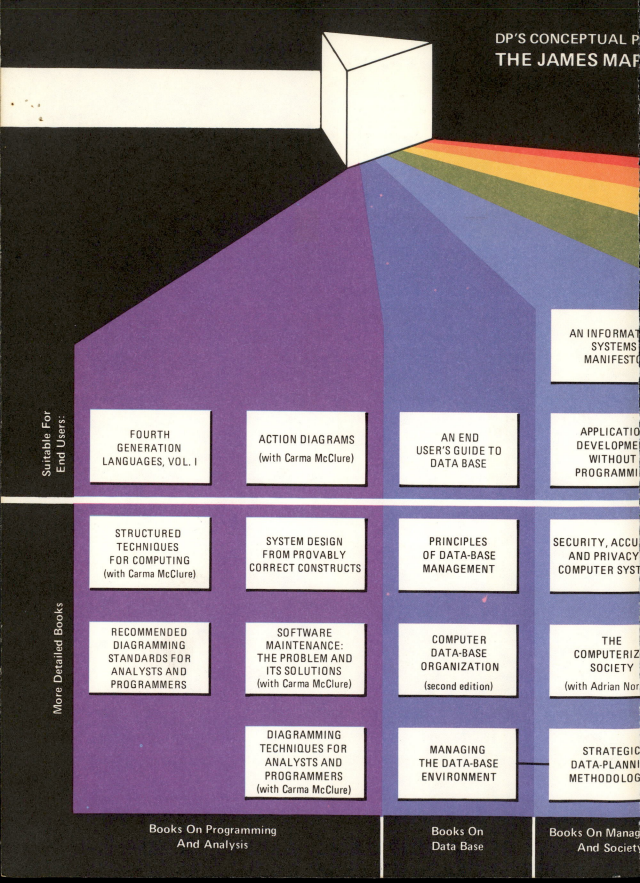